Plant Genetic Resources

An Introduction to their Conservation and Use

Brian Ford-Lloyd and Michael Jackson

Department of Plant Biology
University of Birmingham

Edward Arnold

First published in Great Britain 1986 by
Edward Arnold (Publishers) Ltd, 41 Bedford Square, London WC1B 3DQ

Edward Arnold (Australia) Pty Ltd, 80 Waverley Road, Caulfield East, Victoria 3145, Australia

Edward Arnold, 3 East Read Street, Baltimore, Maryland 21202, U.S.A.

British Library Cataloguing in Publication Data

Ford-Lloyd, Brian
Plant genetic resources: an introduction to their conservation and use.
1. Plant conservation 2. Plant genetics
3. Botany—Variation
I. Title II. Jackson, Michael
639.9 '9 QK86

ISBN 0–7131–2933–6

Test set in Plantin 10/11
by Castlefield Press, Moulton, Northampton
Printed and bound by J. W. Arrowsmith Ltd, Bristol

Preface

Undoubtedly the continuing challenge of the present decade and of those to come, is the demand upon agriculturalists, economists and politicians to feed an ever increasing population. In order to further this objective, plant breeders must develop crops which are capable of yielding well in a variety of environments, and which are resistant to pests and diseases and can grow with reduced fertilizer and pesticide inputs. Sources of genetic diversity thus become increasingly important, and the loss of this diversity in the form of plant genetic resources, is of major international concern.

It is quite clear that the conservation of plant genetic resources, particularly *in situ*, should be viewed as an integral part of any global plan for the conservation of natural resources and ecosystems. However, unlike non-resources, the primary justification for their conservation, provided in this text, lies in their importance in breeding or selecting better varieties and strains of crops for food, fuels and even medicines throughout the world.

We have attempted to show how different sampling and conservation strategies should be applied to different crops. Furthermore, we have aimed at justifying such activities by considering how conserved plant genes have been, and can be, utilized to considerable effect. We have also speculated on how the application of novel biotechnology may enhance the usefulness of plant genetic resources in the future. It has not been possible to include all crops, so we have restricted our discussions principally to the major food crops. However, we have included a discussion on forest genetic resources and other economic and medicinal plants.

International efforts aimed at the exploration and conservation of plant genetic resources have been coordinated since 1974 by the International Board for Plant Genetic Resources. In the past 15 years several specialist books on genetic conservation have been published. We have been strongly influenced by such activities and publications, but most particularly by our close association with Professor Jack Hawkes, one of the pioneers in this area of work, and the instigator of the only postgraduate course on genetic resources, offered by the University of Birmingham, UK. Our own teaching and research in this field, as well as our experience in collecting and evaluating plant genetic resources in different parts of the world have provided us with a useful practical background.

We have been motivated to write this book because of our stimulating association with IBPGR and Birmingham over the past decade or so, and the general need for a text which would serve as an introduction to the field of plant genetic resources, not only for students, but also for scientists and laymen alike, who are perhaps working in other, but related, disciplines.

We are grateful to all colleagues, and their publishers, for permission to include data and figures in this book, which we hope will assist our readers in readily appreciating the importance and complexity of plant genetic resources.

B. V. Ford-Lloyd
M. Jackson

1986

Acronyms used in the text

AVRDC	Asian Vegetable Research and Development Centre, Taiwan
CGIAR	Consultative Group on International Agricultural Research
CIAT	Centro Internacional de Agricultura Tropical, Cali, Colombia (International Centre for Tropical Agriculture)
CIMMYT	Centro Internacional de Mejoramiento de Maíz y Trigo, Mexico City, Mexico (International Centre for the Improvement of Maize and Wheat)
CIP	Centro Internacional de la Papa, Lima, Peru (International Potato Centre)
FAO	Food and Agriculture Organization of the United Nations
IARC	International Agricultural Research Centre
IBP	International Biological Programme
IBPGR	International Board for Plant Genetic Resources, Rome, Italy
ICARDA	International Centre for Agricultural Research in the Dry Areas, Aleppo, Syria
ICRISAT	International Crops Research Institute for the Semi-Arid Tropics, Hyderabad, India
IITA	International Institute of Tropical Agriculture, Ibadan, Nigeria
IRRI	International Rice Research Institute, Los Baños, The Philippines
IUCN	International Union for Conservation of Nature and Natural Resources
TAC	Technical Advisory Committee
UNDP	United Nations Development Programme
USDA	United States Department of Agriculture

Contents

1

The Scope of Genetic Conservation

Current international activities surrounding the genetic resources of plants aim to confront one paradoxical problem. This is that scientists throughout the world are rightly engaged in developing better and higher yielding cultivars of crop plants to be used on increasingly larger scales. But this involves the replacement of the generally variable, lower yielding, locally adapted strains grown traditionally, by the products of modern agriculture — the case of uniformity replacing diversity. It is here that we find the paradox, for these self-same plant breeders are dependent upon the availability of a pool of diverse genetic material for success in their work. They are themselves dependent upon that which they are unwittingly destroying.

Fortunately, this situation has been recognized by a growing number of people, including plant breeders themselves, and concerted efforts are underway to save important plant genes from imminent extinction. In this respect we are dealing with the 'conservation' of plant genes. To most people the word 'conservation' conjures up visions of lovable cuddly animals like giant pandas on the verge of extinction. Or it refers to the prevention of the mass slaughter of endangered whale species, under threat because of human's greed and short-sightedness. Comparatively few people however, are moved to action or financial contribution by the idea of economically important plant genes disappearing from the face of the earth. Precious orchid species with undoubted aesthetic appeal, or the vegetation of an Amazonian rain forest, where sheer vastness cannot fail to impress, may attract deserved attention. But plant genetic resources make little impression on the heart even though their disappearance could herald famine on a greater scale than ever seen before, leading to ultimate world-wide disaster.

Put in these terms, the problem is of some importance. The need for plant gene conservation is highlighted effectively within proposals for a 'World Conservation Strategy' by the International Union for the Conservation of Nature (1980): 'Earth is the only place in the universe known to sustain life. Yet human activities are progressively reducing the planet's life-supporting capacity at a time when rising human numbers and consumption are making increasingly heavy demands on it. The combined destructive impacts of a poor majority struggling to stay alive and an affluent minority consuming most of the world's resources are undermining the very means by which all people can survive and flourish.'

Vanishing resources

Human society, agriculture and earth's abundant plant resources have been co-evolving for more than 10 000 years. Complex interactions have resulted in innumerable patterns of variation, not least in locally adapted plant populations used as food crops, for textiles and for fuels. Artificial selection either unconsciously, or deliberately engineered by humans has meant that diverse wild species have become crops which support the world's population today.

Early on in history, the spread of agriculturally-based communities was accompanied by the movement of plants around the world, with familiar crops being introduced into varied and isolated regions. In the past, primitive cultivation did not inhibit the movement of genes between these crops and their wild relatives. It is only very recently, in line with modern agricultural development and the introduction of new agricultural practices that this gene flow has been reduced, and the evolution of cultivated plants greatly altered or diminished.

The great Russian geneticist and plant breeder, N. I. Vavilov has often been considered as the father of plant genetic resources activities. As long ago as 1926 he proposed that crop plant improvement should draw from wide genetic variation, and to this end he collected cultivated plants and their wild relatives from most parts of the world. These were to provide 'gene pools' from which cultivars could be bred which would be suitably adapted to cultivation in the USSR.

With his extensive studies, Vavilov was able to reveal that genetic variation in cultivated species was concentrated in certain regions of the world which he termed 'centres of diversity' (see Chapter 2). He identified a number of these centres which have been the principal areas from which breeders have collected raw material for their work. In recent years however, the radical changes which have taken place in modern agriculture have presented a formidable threat to the diverse resources held in the Vavilovian centres. These seemingly inexhaustible sources of breeding material have been undergoing a rapid change.

Genetic erosion was already eating into the gene centres of Europe and North America as far back as the 1930s when active plant breeding programmes had been underway for some decades, but the ancient reservoirs of germplasm were still there in more remote parts of the world and seemed to most people to be as 'inexhaustible' as oil in the Middle East. But awareness of a threatening problem came in the sixites, perhaps surprisingly as a result of the 'Green Revolution', when use of high-yielding and genetically uniform cultivars of staple crops spread steadily across the globe. Even in Vavilov's valuable centres, the revolution heralded the abandonment of ancient landraces as well as traditional agricultural practices. Local farmers could hardly be blamed for making every effort to become more efficient in terms of food production. As an example of this, in the last forty years, 95% of the once numerous native varieties of wheat in Greece have been lost forever, the victims of recently introduced commercial varieties.

Hard facts relating to genetic erosion are not easy to come by; what has been lost already can no longer be accounted. One therefore has to resort mainly to personal impressions and subjective accounts. Hawkes, in a report to the International Potato Centre (1973), surveying potato genetic erosion recalls that while taking part in a collecting expedition to Peru and Bolivia in 1971, he was quite aware that 'the richness of varietal diversity had diminished very startlingly, as compared with the situation in 1939' when he had last visited those countries to collect cultivated potatoes. Hawkes defined the cause of this as the work of agronomists and agricultural extension officers who had promoted the cultivation of a limited number of selected varieties, even where new cultivars were not available. Inevitably, fields had been tidied up, with the resulting total absence of wild and weedy forms. While this may be true for increasingly larger areas of cultivation, one of the authors has nevertheless observed that in certain areas of cultivation in Peru, well away from the main centres of population, traditional terrace systems still survive (Fig. 1.1). These still support representatives of ancient cultivated varieties of potatoes, rich in genetic diversity (Jackson *et al.*, 1980).

Fig. 1.1 Ancient Inca terraces at Cuyo-Cuyo in southern Peru. Potatoes and other indigenous tuber crops are grown on these terraces using a traditional farming system. (From Jackson *et al.*, 1980.)

The need to conserve and utilize genetic diversity

Nowadays every major crop that we grow has a very narrow genetic base, and it is this which has contributed to their 'genetic vulnerability'. The problem was highlighted in the United States in 1970 with the southern corn leaf blight epidemic. Massive devastation of crops came about because most of the hybrid varieties in production at that time had a common form of a specific cytoplasmic gene which conferred susceptibility to a particular race of a fungal disease. Equally vulnerable today are sugar beet and sorghum for the same reason. Diverse germplasm of these crops is now being scoured for new sources of cytoplasmic genes.

The history of the potato in Europe again illustrates the necessity for utilizing genetic resources to broaden the genetic base of crop plants. Documented evidence showing that potato breeders in the 19th century were worried about the narrow genetic base of the potato in Europe at that time is available. They used phrases expressing the need for 'new blood' and lamenting the potato's 'degeneration'. It is now fairly certain that all European potatoes existing in the 19th century had been a result of selection over two centuries from a gene pool represented by two initial introductions. It is not surprising therefore that the potato crop was devastated by epidemics of late blight in the 1840s.

A more recent, but less publicized case of genetic vulnerability was caused not by plant disease, but by cold weather. By 1972 in the Soviet Union the famous wheat variety 'Bezostaja' was grown on about 15 million hectares. It had been moved beyond its original area of cultivation far into the Ukraine during a period of relatively mild winters. Then in 1972 a very severe winter occurred, causing losses of millions of tons of winter wheat (Fischbeck, 1981).

The value of local landraces or long-established cultivars and the genes which they may hold, often remains unappreciated until new foreign cultivars are introduced into a region. The semi-dwarf wheats of the Green Revolution when first grown in Mexico were overcome by the fungal diseases black stem rust and stripe rust, while the tried and tested local varieties resisted attack. Hurried crosses had then to be made using genetic resources available locally as one parent and these gave rise to a successful, high-yielding and disease resistant crop. Likewise, upland cotton, introduced from the USA into western Tanzania early this century was severely limited in its productivity by the insect pest cotton jassid, and another disease called bacterial blight. The plant breeder was effectively able to solve these problems when resistance to jassid attack was found to be related to the length and density of hairs on the underside of the leaves (Parnell *et al.*, 1949). In Tanzania, genetic variation for hairiness was found to be present in the locally adapted cottons and was rapidly exploited to give jassid-resistant varieties (Peat and Brown, 1961). Heritable variation for resistance to bacterial blight in these same local cultivars was also exploited (Arnold, 1963) to produce new and highly successful varieties.

The value of local gene pools for the development of regionally adapted cultivars cannot be underestimated. When attempts have been made in West Africa to cultivate pearl millets previously bred in India, they have suffered so much from mildew disease that they could not compete with the less highly bred traditional varieties of local origin. And in Uganda, it was not known that the local cottons possessed a high degree of resistance to *Alternaria* blight until material from nearby Tanzania was introduced into the breeding programme. Despite these examples though, restriction of the use of the germplasm of one region to development of crop varieties in that same region need not apply. For instance, the genetic base represented by potato cultivars in Europe and the United States today can be attributed to the introduction into breeding programmes of genes from many South American primitive varieties and wild species.

By such examples the world community is learning that it can be important to utilize genetic resources for the benefit of the human race.

History of plant genetic resources conservation

In Vavilov's time the idea of active germplasm conservation had generally not been considered. Vavilov's valuable collections were indeed in existence, as were those of the United States Department of Agriculture (USDA) and other governmental agencies. These were being used by breeders to screen for useful characters, but if seed ran out, or was lost, nobody was too bothered. It was thought the material could always be recollected!

There is a hint of an appreciation of the value of true conservation early on when Vavilov, still a young man, was formulating his long-range ideas about species and their systems of variation. Vavilov was clearly influenced by his stay at the then John Innes Horticultural Institution (UK) and by W. Bateson, who in 1923 chaired a Ministry of Agriculture/Royal Horticultural Society committee dealing with fruit trees. One of its objectives was to propose methods of maintaining old commercial varieties of fruit, particularly to conserve the diversity of perennial fruit crops on the principles now attributed to Vavilov.

On an international scale, coordination of activity did not occur until the 1960s. The Food and Agriculture Organization (FAO) of the United Nations spearheaded the initial attack, and organized the first international technical meeting on plant exploration and introduction in 1961. One recommendation of the meeting was to establish a panel of experts to advise and assist FAO. This was placed on a formal footing in 1965, and another panel on Forest Gene Resources was established in 1968. At the same time FAO formed its Crop Ecology and Genetic Resources Unit. During this early period of activity, two further international technical conferences were held under the auspices of FAO and the International Biological Programme (IBP). The first of these, held in 1967, paved the way for subsequent concerted international activity by making certain important recommendations which are worth highlighting as they summarize the major goals set at that time:

(1) the location and nature of genetic resources in the field should be surveyed (i.e. in centres of genetic diversity);
(2) a corresponding survey of material in existing collections should be made;
(3) assembled material should be effectively used and preserved, this being served by adequate classification and evaluation;
(4) strongest emphasis should be placed upon the conservation of genetic resources;
(5) efficient documentation should be carried out at all stages of activity;
(6) international coordination, guidance and administrative backing should be sought at the highest level.

It was further urgently stressed that unequivocal leadership should be assumed if the survival of some of man's most vital resources was to be ensured.

Perhaps with these persuasive words, the recently-established Consultative Group on International Agricultural Research (CGIAR) and its Technical Advisory Committee (TAC) convened a working group of world-renowned scientists in 1972 which recommended the creation of a network of nine regional genetic resources centres and a series of crop-specific centres. The U.N. Conference on the Human Environment in 1972 also took up the cause and adopted a resolution calling for an international programme for preserving germplasm of tropical and sub-tropical crops. However, it was not until the International Board for Plant Genetic Resources (IBPGR) was established by the CGIAR in 1974 that real work began.

IBPGR now works in close association with FAO, and its primary function lies in the development and administration of the international plant genetic resources conservation network. In the years since it was created, a great deal of progress has been made towards the organization of a global network of genetic resources centres, and a large number of collecting missions has been carried out. IBPGR has had a dramatic catalytic effect upon conservation efforts of scientists and agricultural centres throughout the world.

Amongst the Board's major tasks are the promotion, collection, conservation and evaluation of plant genetic resources of species of major economic importance and their wild relatives. It has defined over 50 priority crops needing urgent action and priority regions requiring survey (Table 1.1). The Board has aimed at the establishment of standards, methods and procedures for exploration, evaluation and for conservation of stocks of both seeds and vegetative material, to be applied particularly in the network of replicated storage centres or genebanks (Fig. 1.2). The aim has been to promote the dissemination of both information and genetic material among these centres; this has been supported by research into effective information storage and retrieval systems.

Table 1.1 Priority crops needing urgent action. (From IBPGR, 1983a).

Crop	Global priority 1	Global priority 2	High regional priority■
Cereals	Wheat	* Sorghum * Finger millet * Barley * Pearl millet * Foxtail millet * Rice	Maize Quinoa
Food legumes	*Phaseolus* beans	* Groundnut * Soyabean * Cowpea * Yard long bean * Winged bean * Chickpea * *Vigna radiata* *V. mungo* *V. aconitifolia* *V. umbellata*	*Vicia faba* Lentil Lupin

Table 1.1 continued

Crop	**Global priority 1**	**Global priority 2**	**High regional priority■**
Roots and tubers	Cassava Sweet potato	Potato	Yam Taro and Aroids Minor S. American tubers
Oil crops		Oil palm (*Elaeis melanococca*) * Coconut * Oilseed brassicas	
Fibres		Cotton	
Starchy fruits		* Starchy banana and Plantain	Breadfruit and Jackfruit
Sugar crops		* Beet * Sugarcane	
Beverages	Coffee	Cocoa (* *Criollo* varieties)	
Subtropical and Tropical fruits		* Dessert banana * Citrus * Mango	Avocado *Lansium* *Annona* *Passiflora* Peach palm Durian Rambutan
Temperate fruits		* Apple * Pear and Quince Peach and Nectarine	
Vegetables	Tomato	* Amaranth * Brassica * Cucurbits * Eggplant * Okra * Onion * Chilli * Radish	Bitter gourd Globe artichoke *Cucumis* *Sechium* Kangkong *Spinacia*
Trees		Trees for fuelwood and environmental stabilization	

* = a priority in at least one region.
■ Although having a lower global priority, these crops all have a first priority in at least one region.

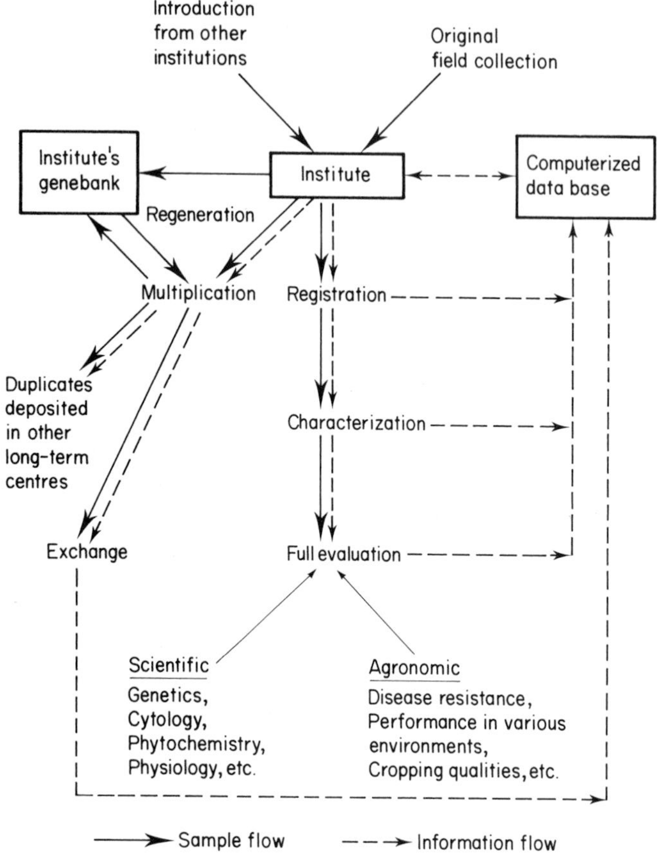

Fig. 1.2 Procedures in evaluation and handling of accessions in genebanks. (From CGIAR, 1982.)

IBPGR has also been highly active in supporting training of personnel capable of carrying out the various tasks involved in the whole genetic resources system. The University of Birmingham in the United Kingdom initiated the first International Training Course in Genetic Conservation leading to the M.Sc. degree. This course provides a broad theoretical and practical background to all aspects of a genetic conservation programme. The Board also supports a number of short technical courses which cover many practical aspects of genetic resources work and are held in different locations throughout the world for various lengths of time.

Apart from the immediate conservation activities, above all else in importance is the Board's effort to stimulate the availability of genetic material for plant breeding and other scientific activities. It is after all, the practical utilization of genetic resources in crop development that is the justification for the whole operation.

Feeding the world

At the start of this decade between 10 000 and 12 000 people were dying of hunger every day. One person in five was undernourished and one in ten seriously so — including 100 million children under the age of five. This has been the product of the population explosion of the latter half of the 20th century, which has also brought with it not only pressure on agricultural production, but natural resources, particularly in developing countries where 80% of the enormous increase in population is taking place — Africa, Latin America and parts of Asia. At the same time the demand from the developed and developing countries for more and more of the world's natural resources has caused depletion of forests, savannah and scrub land, the destruction of ecosystems by forest felling, the drainage of marshes and the general replacement of the natural vegetation by arable farming and pasturing. This is the result of bringing more land under the plough in order to increase food production. But since much land thus taken for agriculture cannot really support it, the results have not always been successful. This has contributed to the dependence on subsistence agriculture, where annual output is often barely enough to support the village or family unit, with little or no surplus to generate cash. Even where conditions are more favourable, the urban population is usually increasing faster than the locally available surplus.

In the short-term, therefore, widespread famine can only be averted by raising crop yields, particularly in the developed nations, in order to sustain supplies of direct food-aid. This will be done by improving production through more efficient farming methods: the greater applications of fertilizers, pesticides and insecticides. This has already proved successful in the European Economic Community, to the extent that we are now hypercritical of overproduction leading to food 'mountains'. Our criticism should really be aimed at an economic system which is incapable of redistributing food supplies to the hungry mouths in the Third World. Nevertheless, much can be done in developing countries themselves by improving farming practices in a more basic way, and particularly through the application of efficient irrigation schemes in areas where rainfall is insufficient.

But in the long run, the chronic problems of undernourishment and malnutrition can be best tackled by a combination of better farming and better varieties — by introducing varieties of cultivated plants that can make best use of irrigation and other farming inputs.

Today the world population is around 5000 million; the figures may be 6500 million by 2000 AD and could reach between 8000 and 15 000 million next century. Will the politicians, agronomists and plant breeders have the answer? One fact is indisputable: the plant breeders will not be able to provide any panacea without gene resources to work with.

2

Crop Evolution and Diversity

An understanding of the nature and extent of crop plant diversity is fundamental to plant genetic resources. Indeed it could be argued that without this we would be unable satisfactorily to collect, conserve or utilize the vast array of diversity which is found in the majority of crop plants and their wild relatives. Crop plants have evolved from wild species under the influence of humans. The relationships between crops and their wild relatives have been determined for many important crop species, but for others, their origins and evolution under domestication are shrouded in mystery. From the practical point of view, the identification of the wild progenitors of crops is of great significance, and such studies provide valuable information for the plant breeder, as will be shown in Chapter 6.

In order to understand the evolution of crop plants, a discussion of the origins of agriculture, the time and place of origin, and the causes and patterns of crop plant diversity is relevant here, and provides a framework for a more detailed discussion of several of our major food crops.

The origins of agriculture

When the available evidence for the origins of agriculture is examined there is a fair degree of concordance between different workers as to how, where and when humans first began to domesticate plants and to cultivate them. What will remain a mystery, perhaps, is why agriculture began as we believe it did some 10 000 years ago in various parts of the world. Various hypotheses have been put forward to explain this, and surely no single idea is correct. At the end of the last Pleistocene glacial period the climate began to ameliorate, in areas first that were closer to the Equator, and as the ice sheets retreated, conditions favourable for plant growth became more widespread. Undoubtedly the human population expanded, and it has been suggested that humans only reached a sufficient cultural level then. Were the beginnings of agriculture the result of conscious experimentation by our ancestors, or were they, as Hawkes (1983) has put it 'a stroke of inspiration'? Whatever the reason, the advent of the so-called 'Neolithic Revolution' and the domestication of plants was a major event in man's cultural evolution.

Until the Neolithic Revolution, which was probably a slow process rather than the dramatic event which the term 'revolution' implies, pre-agricultural man was an exploiter of natural ecosystems. Harris (1969) has recognized two types of pre-agricultural subsistence economies, the specialized hunters whose subsistence depended upon the intensive exploitation of relatively few wild species, and the generalized gatherer-hunter-fisher populations who exploited less intensively a broader spectrum of food resources. It is these latter populations which Harris suggests were more likely to have initiated plant domestication, because of their less nomadic existence and their familiarity with the plants which surrounded them.

It is convenient to use a conceptual framework of an ecological approach to agricultural origins and plant domestication, and this is a theme which we shall return to later. In the context of this book, we shall be less concerned with why agriculture began when it did than with the consequences of the initiation of plant domestication and what plants were first cultivated.

The origins of agriculture and the subsequent sequence of crop evolution can be placed in a time framework by using the extensive archaeological evidence which is available. In fact we are in a position with crop plants to study the evolutionary changes which have taken place over a period of about 10–15 000 years, something which we cannot do with wild plants, because of the lack of an extensive fossil record. With crop plants and their wild ancestors, many sites provide abundant evidence from remains of the plants themselves which had been discarded around human settlements.

Let us now turn to the areas in which agriculture developed. Both De Candolle (1882) and Vavilov (1926) pointed out that the domestication of our crops seems to have taken place in several distinct and well-defined areas of the world, mainly within the tropics and sub-tropics, and generally, though not always, in mountainous regions. From the archaeological evidence it seems clear that agriculture originated independently in several areas, most notably South West Asia, in Meso-America and in China at more or less the same time. Interestingly, our most important crops were domesticated from a rather restricted number of flowering plant families, the leading ones being the Gramineae, Leguminosae, Solanaceae, Cruciferae and the Rosaceae, but this is not to say that other families have not also been exploited by humans through gathering activities.

What were the characteristics of the ancestors of our crop plants which made them of value to humans? Hawkes (1983) has suggested that if we look at agricultural origins from an ecological point of view, it is possible to draw several valid conclusions. Obviously the actual ancestors of crop plants cannot be examined — only present-day wild relatives and primitive cultivated forms can be studied. If we compare these two types of plants there is one feature which is common to both, and this is their weediness. Most domesticated herbaceous plants and their wild relatives are ecological weeds, that is, plants with the ability to colonize open or disturbed habitats with bare soil, and which are unable to withstand a high level of competition from other plants. Where would these plants have existed before their exploitation by humans? Such plants tend to be colonizers of natural bare lands such as riverbanks, gravel roads, landslides and estuarine plains.

Did humans adopt a settled mode of existence as a result of plant domestication, or did domestication reinforce such settlements? Even nomadic peoples create limited settlements, and in these areas, the land becomes disturbed, and with the addition of organic refuse and human waste, the conditions which such ecological weeds favour are provided in close proximity to humans. One hypothesis for the origin of agriculture is the 'rubbish heap' hypothesis first expounded by Engelbrecht (1916), and later developed by the American geneticist and student of crop origins Edgar Anderson (1969). This rubbish heap hypothesis suggests that ecological weeds colonized the areas around human dwellings, were gathered and eventually brought into cultivation. Some of these plants were most likely those which had been gathered, and from which seeds and other propagules had been discarded. In this way, the intimate relationship between man and these plants was reinforced. On the other hand, many of these ecological weeds must have been unpalatable and were never domesticated.

The origins of seed agriculture and root and tuber agriculture, or vegeculture, as it is sometimes referred to, are in essence similar, although there are some differences. Whether one developed from the other will never be known, although Sauer (1952) has argued that vegeculture originated first. Seed agriculture appears to represent the indigenous mode of agriculture in the drier tropics and sub-tropics of both the Old and New Worlds. Vegeculture is more typical of the humid tropical lowlands of America and South East Asia.

The archaeobotanical evidence for the origins of cereals and grain legumes points to mountainous areas in both the Old and New Worlds, which have marked wet and dry seasons. Under these climatic conditions, ecological weeds would need to germinate and grow quickly when the rains came in autumn and spring, and to complete their reproductive cycles before the high temperatures of summer baked the soil. Annual plants would be at a selective advantage if they had large food reserves which enabled them to survive drought and also to grow quickly once the rains came. In this sense, these plants would be pre-adapted for agriculture and the two factors of 'weediness' and large food reserves do seem to be the key to the domestication of Old World mountain seed crops (Hawkes, 1969).

Hawkes (1969) has also recognized three stages for the development of seed agriculture, namely (1) gathering and colonization; (2) harvesting; and (3) sowing or planting. These stages can also be ascribed to the development of vegeculture. Gathering and colonization represents that stage when ecological weeds began to colonize the areas around human settlements. As will be seen in a later section (p. 13), the domestication of plants resulted in a number of fundamental changes in the plants themselves. What is clear is that during the process of domestication, plants lost the ability to survive as wild plants, as mutations which enabled man to harvest plants more easily, such as the non-brittle rachis mutation in cereals, or the loss of legume dehiscence in the pulses, arose in plant populations. The process of harvesting actively favours these mutations, but such plants would be at a considerable disadvantage in the wild. Undoubtedly similar mutations do arise at each generation, and although in themselves are not deleterious to wild plants, it is clear that when all seeds fall to the ground together, severe inter-plant competition results, and these plants will succeed in producing fewer and fewer plants for subsequent generations, relative to plants which are not at this selective disadvantage.

In the early development of the harvesting stage, Hawkes (1969) has suggested that all non-brittle and non-dehiscent plants were consumed, and susequent crops were almost entirely produced from wild type forms. Consequently there must have been considerable selection pressure against these mutant forms, which ultimately revolutionized the characteristics of these plants. When, as eventually happened, man began to retain some of the seeds for sowing, selection pressures changed in favour of the non-brittle mutants. Only at this stage could crops be considered to be domesticated.

Vegeculture must have originated in tropical areas with a marked wet and dry season, because it is under these conditions that root and tuber crops would have evolved large food reserves in their underground organs. Large reserves would enable plants to grow quickly at the beginning of the rainy season. Various authors have concluded that vegeculture originated at the edges of dry forest zones, at low altitudes, or at least in the contact zones or ecotones between major ecosystems, such as forest and savannah (Sauer, 1952; Harris, 1969).

Mountainous areas provide conditions for the rapid differentiation of races and varieties of different crops. In these areas, there is a wealth of micro-climates, of isolated valleys, and under such conditions wild and cultivated plants may evolve rapidly (Vavilov, 1926). Vavilov also commented on the ethnic diversity of South West Asia, of the Caucasus and North West India, and in these areas many of our crops were first domesticated.

The evolutionary dynamics of plant domestication

Domesticated plants are adapted to permanently man-made habitats which have been specially created for them by man. In fact, de Wet and Harlan (1975) and de Wet (1979) have suggested that plant domestication is evolution in a man-made habitat. Crops have evolved under domestication to the point where they depend on humans both for habitat and propagation.

The process of domestication has resulted in a number of phenotypic changes as well as in reproductive biology, characters which make the plants more acceptable to humans. Apart from the obvious feature of the loss of a dispersal mechanism such as the non-brittle rachis mutant in cereals, others are often characteristic of domesticates rather than wild plants. Crop plants often show gigantism, especially in the parts of the plant which have been selected by man, such as seeds and fruits. This increase in size is one of the criteria used in the examination of archaeobotanical remains.

Crop plants are subjected to the rigours of both natural and artificial selection pressures. Wild plants must be adapted to the environment in which they grow, and plants which diverge significantly may well be selected against. In many ways, cultivated plants have been released from some of these selection pressures, and great diversity, particularly morphological diversity, has accumulated in many crops. Thousands of cultivars of the important cereals have been recognized; potato cultivars are represented by a wealth of tuber shapes and colours, and the same can be said of the fruits of cucurbits or the

Fig. 2.1 Morphological variation in potatoes, beet and cucurbits (*left*), and a wheat landrace in Algeria (*above*).

roots of beets for example, and in fact of almost any cultivated plant one can imagine (Fig. 2.1).

During domestication, crop plants were brought into contact with many different wild plants, with which they hybridized and exchanged genes. In doing so, the physiological adaptation of the plants was increased, as is exemplified by bread wheat, which combines three genomes from diploid wheat and two species of goat-grass or *Aegilops*. As a consequence, the adaptive ability of wheat was dramatically increased, with the result that it became one of the most important of our crop plants.

Other features which have changed through the process of domestication include the suppression of defensive mechanisms, the reduction of sexual fertility in vegetatively-propagated crops, and a change of habit, including a change from the perennial to the annual habit. A change in reproductive biology from outbreeding to inbreeding has enabled crop plants to exploit fully the man-made habitat. In areas with a Mediterranean climate, there is a wealth of microclimates within even small areas. In order for long-term survival, it is advantageous for plants to produce adapted progeny through inbreeding. Several of the wild relatives of the cereals and pulses are also inbreeders. In Israel, for example, they are confined to the open habitats favoured by colonizing species or ecological weeds.

Several morphological features are associated with the change to inbreeding and domestication, and some of the cereals are even cleistogamous. Pollen to ovule ratios are lower in inbreeding species, a phenomenon widespread in flowering plants (Cruden, 1977), and bearing an evolutionary significance which has been assessed by Uma Shaanker and Ganeshaiah (1980). Reproductive changes in finger millet have been documented for example (Ganeshaiah and Uma Shaanker, 1982) including synchronicity of flowering, and an increase in gynoecium volume. These and other features are discussed in a general account of cereal domestication given by De Wet (1975).

Another aspect of importance of evolution in man-made habitats is the formation of weed races. Domesticated species are almost always characterized by wild, weed and cultivated races, and this is particularly characteristic of cereals. Weed races of cereals commonly originate as derivatives of hybrids between wild and cultivated races, and often accompany the crop beyond its natural distribution. The genetic links between cultigens and their associated weed races have been studied. Good examples are sorghum and pearl millet (*Pennisetum americanum*). Hybrids between wild and cultivated forms are often fully fertile, and although some gene exchange does occur through introgressive hybridization, this is not extensive because of the action of disruptive selection. Weed races mimic the cultivated race they accompany in vegetative and inflorescence characteristics, but unlike the cultigens they do have an efficient seed dispersal mechanism. Weed races may have diverged simultaneously from a shared wild ancestor. Whatever their mode of formation, these 'crop-weed complexes' as they are called, may considerably affect the amount of genetic diversity in a crop, depending upon the phylogenetic relationships and the extent of gene exchange between the crop and weed (Pickersgill, 1981). For this reason, in terms of plant genetic conservation, the importance of weed races cannot be emphasized too much.

The time and place of the origin of agriculture

Students of crop plant evolution are extremely fortunate that they can, with a fair degree of certainty, place the origins of agriculture and the domestication of plants both in time and geographically. Advances in archaeological research, and the discovery of further sites in various parts of the world have all contributed. Primitive humans were extremely untidy, and archaeological deposits testify to this, with organic remains frequent in many sites, as well as remains of pottery, tools and textiles, all of which tell us something about our forefathers.

But what of the archaeological evidence itself? In many sites there are plant remains, particularly in the arid and semi-arid areas, which have become carbonized, but which still retain many of their gross morphological features. Comparison with present-day plants allows us to tell whether the remains are from wild plants, and therefore represent a pre-agricultural culture or at least incipient agriculture, or whether the remains come from domesticated plants.

There is surprising concordance amongst researchers as to where agriculture originated, the so-called 'cradles of agriculture'. From these cradles, agriculture spread to other parts of the world. There is considerable archaeological evidence for incipient agriculture in Thailand about 13 000 years ago (Gorman, 1969), in the Near East about 11 000 years ago (Cambel and Braidwood, 1970) and in Mexico some 8000 years ago (MacNeish, 1964). These sites hold some of the earliest evidence for early agricultural development. There are indications that agriculture had an independent origin in China (Ho, 1969) based on the cultivation of millets, but agriculture may have spread there from Thailand, as it probably did to Europe and parts of Africa from the Near East and S.W. Asia. There are also early sites in Peru, particularly on the coast, which have been dated as early as 7–8000 BP (Before

Present), with evidence of agriculture based on the cultivation of squashes and root crops such as cassava and sweet potato.

While the evidence for the origins of seed agriculture is impressive, that for vegeculture is far less certain. Fruits and seeds are well preserved in the arid and semi-arid areas particularly of the Near East, where the story of the origins of cereals and grain legumes has been deciphered to the greatest extent. Under the hot, humid conditions of the tropical lowlands, organic remains of roots and tubers are rapidly decomposed. Consequently archaeological evidence is poor, and often only of tools and other equipment, which suggest the utilization of plants which may have been collected, and not actually cultivated.

Patterns of crop diversity

Alphonse de Candolle made the first significant contribution to our understanding of agricultural origins in his book *Origine des Plantes Cultivées*, published in 1882. But it was Vavilov who undoubtedly focussed attention on the diversity to be found in crop plants, and to the fact that it was concentrated in what he termed 'centres of diversity' which he interpreted as centres of origin for many crops. He investigated the distribution of the major crop plants, and determined areas where there were concentrations of botanical varieties, using detailed studies of their morphology, cytology, genetics and resistance to pests and diseases, and adaptation to different climatic conditions. As a result, Vavilov recognized eight centres of origin, shown in Fig. 2.2, and an indication of their crop diversity is given in Table 2.1. These centres lie between 20° and 45° latitude north and south of the equator, in mountainous regions. Vavilov assumed that agriculture developed independently in these areas, because of differences in agricultural method, implements and domestic animals.

Table 2.1 World centres of diversity (centres of origin *sensu* Vavilov) of cultivated plants. (From Zohary, 1970.)

1. THE CHINESE CENTRE	2. THE INDIAN CENTRE
Avena nuda, Naked oat (secondary centre of origin)	*Oryza sativa*, Rice
Glycine max, Soybean	*Eleusine coracana*, Finger millet
Phaseolus angularis, Adzuki bean	*Cicer arietinum*, Chickpea
Phaseolus vulgaris, Bean (recessive form; secondary centre)	*Phaseolus aconitifolius*, Math bean
Phyllostachys spp., Small bamboos	*Phaseolus calcaratus*, Rice bean
Brassica juncea, Leaf mustard (secondary centre of origin)	*Dolichos biflorus*, Horse gram
Prunus armeniaca, Apricot	*Vigna sinensis*, Asparagus bean
Prunus persica, Peach	*Solanum melongena*, Egg plant
Citrus sinensis, Orange	*Raphanus caudatus*, Rat's tail radish
Sesamum indicum, Sesame (endemic group of dwarf varieties; secondary centre)	*Colocasia antiquorum*, Taro yam
Camellia (Thea) sinensis, China tea	*Cucumis sativus*, Cucumber
	Gossypium arboreum, Tree cotton, 2x
	Corchorus olitorius, Jute
	Piper nigrum, Pepper
	Indigofera tinctoria, Indigo

continued

Table 2.1 continued

2a. THE INDO-MALAYAN CENTRE
Dioscorea spp., Yam
Citrus maxima, Pomelo
Musa spp., Banana
Cocos nucifera, Coconut

3. THE CENTRAL ASIATIC CENTRE
Triticum aestivum, Bread wheat
Triticum compactum, Club wheat
T. sphaerococcum, Shot wheat
Secale cereale, Rye (secondary centre)
Pisum sativum, Pea
Lens culinaris, Lentil
Cicer arietinum, Chickpea
Sesamum indicum, Sesame (one of the centres of origin)
Linum usitatissimum, Flax (one of the centres of origin)
Carthamus tinctorius, Safflower (one of the centres of origin)
Daucus carota, Carrot (basic centre of Asiatic varieties)
Raphanus sativus, Radish (one of the centres of origin)
Pyrus communis, Pear
Malus pumila, Apple
Juglans regia, Walnut

4. THE NEAR EASTERN CENTRE
Triticum monococcum, Einkorn wheat
Triticum durum, Durum wheat
Triticum turgidum, Poulard wheat
Triticum aestivum, Bread wheat (endemic awnless group; one of the centres of origin)
Hordeum vulgare, Endemic groups of cultivated two-rowed barleys
Secale cereale, Rye
Avena byzantina, Red oat
Cicer arietinum, Chickpea (secondary centre)
Lens culinaris, Lentil (a large endemic group of varieties)
Pisum sativum, Pea (a large endemic group; secondary centre)
Medicago sativa, Blue alfalfa
Sesamum indicum, Sesame (a separate geographic group)
Linum usitatissimum, Flax (many endemic varieties)
Cucumis melo, Melon
Prunus amygdalus, Almond
Ficus carica, Fig
Punica granatum, Pomegranate
Vitis vinifera, Grape
Prunus armeniaca, Apricot (one of centres of origin)
Pistacia vera, Pistachio (one of the centres)

5. THE MEDITERRANEAN CENTRE
Triticum durum, Durum wheat
Avena strigosa, Hulled oats
Vicia faba, Broad bean
Brassica oleracea, Cabbage
Olea europea, Olive
Lactuca sativa, Lettuce

6. THE ABYSSINIAN CENTRE
Triticum durum, Durum wheat (an amazing wealth of forms)
Triticum turgidum, Poulard wheat (an exceptional wealth of forms)
Triticum dicoccum, Emmer
Hordeum vulgare, Barley (an exceptional diversity of forms)
Cicer arietinum, Chickpea (a centre)
Lens culinaris, Lentil (a centre)
Eragrostis tef, Teff
Eleusine coracana, Finger millet
Pisum sativum, Pea (one of the centres)
Linum usitatissimum, Flax (a centre)
Sesamum indicum, Sesame (basic centre)
Ricinus communis, Castor bean (a centre)
Coffea arabica, Coffee

7. THE SOUTH MEXICAN AND CENTRAL AMERICAN CENTRE
Zea mays, Corn
Phaseolus vulgaris, Common bean
Capsicum annuum, Pepper
Gossypium hirsutum, Upland cotton
Agave sisalana, Sisal hemp
Cucurbita spp. Squash, Pumpkin, Gourd

8. SOUTH AMERICAN (PERUVIAN-ECUADOREAN-BOLIVIAN) CENTRE
Ipomoea batatas, Sweet potato
Solanum tuberosum, Potato
Phaseolus lunatus, Lima bean
Lycopersicon esculentum, Tomato
Gossypium barbadense, Sea island cotton (4x)
Carica papaya, Papaya
Nicotiana tabacum, Tobacco

8a. THE CHILOE CENTRE
Solanum tuberosum, Potato

8b. BRAZILIAN-PARAGUAYAN CENTRE
Manihot esculenta, Manioc
Arachis hypogaea, Peanut
Theobroma cacao, Cacao (secondary centre)
Hevea brasiliensis, Rubber tree
Ananas comosa, Pineapple
Passiflora edulis, Purple granadilla

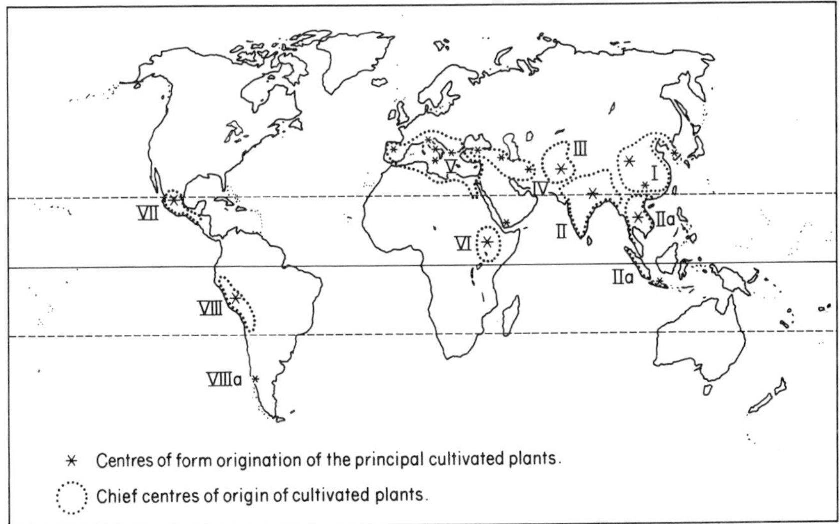

Fig. 2.2 Vavilov's world centres of origin of cultivated plants (Vavilov, 1951). Stars indicate centres of form origination of the principal cultivated plants; dotted areas indicate chief centres of origin of cultivated plants. (From Hawkes, 1983.)

It is probable that Vavilov based his ideas on the 'Age and Area' hypothesis of Willis (1922), which briefly states that the longer an organism has been present in an area, the more diverse it will be. Vavilov made a profound contribution to our understanding of agricultural origins and the geographical distribution of genetic diversity in crop plants. Today, however, his ideas have been modified. Zohary (1970) has indicated that although the reality of centres of diversity in certain parts of the world can be demonstrated, calling them 'centres of origin' is only an interpretation of the biological facts.

In looking at Vavilov's gene centres, notable gaps are Africa (except for Ethiopia), North America and Australia. The situation for North America and Australia can be easily explained as few crops have been domesticated in these continents. Examples from North America are the sunflower (*Helianthus annuus*) and the Jerusalem artichoke (*Helianthus tuberosus*). Vavilov did not visit West Africa, and consequently this was not included in his scheme. It is now known that West Africa was an important cradle of agriculture, with the development of cultures based on the cultivation of African rice, *Oryza glaberrima*, and yams, *Dioscorea* spp. Since Vavilov's work several researchers have proposed different schemes, including Darlington and Janaki-Ammal (1945), Portères (1950) for tropical Africa, Harlan (1951) with his concept of gene microcentres, Murdock (1959) who established four regional agricultural complexes, and culminating in the megacentres of Zhukovsky (1968), a scheme which covered practically the whole globe such that 'centres' could not be recognized. Harlan (1971a) developed the idea of centres and non-centres, and suggested that agriculture began independently in three areas, each of these centres associated with a large and diffuse 'non-centre'. So we have a Near East centre and an African non-centre, a Chinese centre and its associated South East Asian and Pacific non-centre, and a Meso-American centre and a South American non-centre. A comprehensive review of these ideas is given by Zeven and de Wet (1982).

Recently, Jack Hawkes (1983) has put forward an alternative scheme, in which he distinguishes nuclear centres of agricultural origins from the regions of diversity which developed later when farming had spread out from the nuclear centres in which it had originated. Hawkes' scheme is shown in Table 2.2.

Table 2.2 Nuclear centres and regions of diversity of domesticated plants. (From Hawkes, 1983.)

Nuclear centres	*Regions of diversity*	*Outlying minor centres*
A Northern China	I China	1 Japan
	II India	2 New Guinea
	III Southeast Asia	3 Solomon Islands, Fiji, and South Pacific
B The Near East	IV Central Asia	4 Northwestern Europe
	V The Near East	
	VI The Mediterranean	
	VII Ethiopia	
	VIII West Africa	
C Southern Mexico	IX Meso-America	5 United States, Canada
		6 The Caribbean
D Central to southern Peru	X Northern Andes (Venezuela to Bolivia)	7 Southern Chile
		8 Brazil

Whichever scheme is correct is perhaps irrelevant in one sense. What is important is that diversity in crop plants, in the form of plant genetic resources, is more concentrated in certain regions than in others, and it is to these that genetic conservationists must turn. Vavilov's contention that 'centres of origin' are characterized by dominant alleles and that the frequency of recessive alleles increased and diversity decreased towards the periphery as a result of inbreeding, geographical isolation and drift, has been challenged by Yamashita (1979) and Witcombe and Gilani (1979). They have suggested that perhaps diversity has been increased in peripheral areas, through the release of variation in ecologically diverse environments. What is also certain is that Vavilov failed to recognize centres such as Ethiopia and Central Asia as 'accumulation centres', which were proposed by the German scientist Elizabeth Schiemann. Vavilov had claimed that these two were the centres of origin for durum and bread wheats, respectively. We are now aware that there are no wild wheats in Ethiopia, and that bread wheat has no hexaploid wild progenitor, and that these wheats must have been taken to these areas as domesticated plants. Variation was released, and diversity intensified following cultivation under the varied ecological and agricultural conditions there. In addition, it is also known that crops such as rye and oats have centres of diversity in the Near East, but that they in fact developed as 'secondary crops' within primary crops such as wheat and barley. Rye and oats were carried along as weeds with wheat and barley, west and northwards into Europe. As conditions became less favourable for the primary crops, then the proportion of the 'weeds' increased, until eventually the weed became the crop and the crop the weed, and wheat and barley were replaced.

From his observations of crop diversity, Vavilov also noted that similar variation patterns were often seen in two or more unrelated crops in a given geographical area. His Law of Homologous Series was proposed to systematize such examples of parallelism (Vavilov, 1951). These occur because crops have most likely responded to similar natural and artificial selection pressures. For example, different cereals such as wheat and rye have been found in the same area with similar spike characteristics, and unrelated potato species from Mexico have been found with resistance to the late blight fungus (Hawkes, 1983). The importance of Vavilov's law is that it has predictive value, in that if one crop plant is found with a particular trait, then this same trait is likely to be found in unrelated species from the same area.

The genetic variability of crop plants

The pattern of genetic diversity seen in crop plants results from the interaction of five important factors; namely gene mutation and migration, recombination, selection and genetic drift. The first three of these factors all enhance variation in natural populations, whereas selection and drift tend to act to reduce variation. Their relative importance in crop plants can be assessed from the following discussion.

Mutation is a process by which novel variation is formed, and whereby new characters are found in populations. Mutation is a spontaneous phenomenon and occurs at a certain rate for each locus in each organism, varying between 10^{-3} and 10^{-6}. There is also a mutation pressure on each generation, and there is no reason to believe that the mutation rate is different in wild populations and in evolution under domestication. Migration can increase variation in populations by adding allied genetic material, but this is hard to demonstrate in populations of wild plants. However, this phenomenon is very important for evolution under domestication. During their wanderings, humans have taken their crops with them to new sites, and geographical variation has been increased through adaptation to different environmental conditions. Hybridization with allied species, not found in the original region of domestication has often occurred, with stablization of hybrids through polyploidization endowing the progeny with additional variation and adaptation. A good example is bread wheat, *Triticum aestivum*, which probably originated in an area around the southern end of the Caspian Sea. One of the parents of bread wheat, *T. dicoccum* or emmer, was domesticated further west, but was carried eastwards until it came into contact with the wild goat-grass, *Aegilops squarrosa*, and hybrids were formed, resulting in the formation of hexaploid bread wheat.

Recombination is the mechanism by which variation is arranged in different ways. Here the reproductive biology of the plants, whether cross-pollinating or inbreeding, should be considered. Variation in outbreeders is greater because of heterozygosity at different loci, but in inbreeders homozygosity reaches a high level. In some outbreeding crops, various mechanisms such as structural hybridity have evolved to reduce recombination. The importance of this lies in the fact that often desirable gene combinations, or what might be called adaptive gene complexes, have evolved. Again there is no reason to believe that the process of recombination is different in wild and cultivated forms, except that under domestication the mode of pollination does often change, from one of cross-pollination to self-pollination. Why has this occurred so frequently? Perhaps one explanation can be found with the example of the tomato. Wild tomatoes are out-pollinators relying upon bees for effecting pollination. There is considerable evidence to indicate that the tomato was domesticated in Mexico, whereas the diversity of wild species occurs in South America. Pollinators may not be present in areas where cultigens are formed, so that there is a very strong selection pressure for seed production through selfing.

Those factors which reduce variation will now be considered. Selection is the process by which undesirable genes or gene combinations are eliminated from populations. A successful gene combination will reproduce more frequently, so there is some degree of differential fertility in populations between different genotypes. Genes which place an individual at some sort of disadvantage will be selected against and individuals with these characteristics eliminated. Both natural and artificial selection pressures are operative, but the selection pressures acting on wild and cultivated populations are very different and are sometimes opposite to one another. Whereas characteristics such as lack of seed dispersal and uniform germination are often selectively disadvantageous in wild plants, humans favoured these, albeit unconsciously in the early stages of domestication.

The last factor to consider is genetic drift, and it covers two important concepts; firstly the random change in genetic equilibrium in plant populations, especially small populations, and secondly the founder principle which is an important feature of crop evolution. The small amount of variation in a population may have resulted from the founding of the mother population by a small number of individuals. What is the significance of drift in domestication of crop plants? Again, the cereals can be used as an example. When the non-brittle rachis mutant developed, with the loss of the seed dispersal mechanism, this must have been represented in only a very small proportion of the wild progenitor population.

A discussion of the founder principle leads on to a consideration of the formation of polyploids and their significance in crop plant evolution. Allopolyploids, formed following hybridization between species with distinct genomes, as has happened in wheat, cotton, bananas and many other crops, represent the progeny of probably only a few hybridizations. Once hybrids have been formed, it is important for meiosis to be stabilized and fertility restored in sexual species, and this can be achieved through polyploidization. In vegetatively reproducing crops, the restoration of seed fertility may not be so important for the survival of hybrids, but for seed crops, unless fertility is restored, they will be eliminated quickly from the population. Indeed we are now aware that often chromosome pairing within genomes is controlled genetically, and without this control allosyndetic pairing may occur. This phenomenon has perhaps been best illustrated in wheat, in which a gene preventing homoeologous chromosome pairing is located on chromosome 5B.

But what of the evolutionary significance of hybridization and polyploidy? Harlan (1966) considered that the evolution of crop plants progresses primarily on a differentiation-hybridization cyclic system. Hybridization between two varieties of a domesticated plant presupposes a period of selection that led to the differentiation of the two varieties. The length of the differentiation-hybridization cycle is simply an index of the degree of divergence that can take place without preventing hybridization between forms, and it depends upon the extent to which a genotype of a crop is buffered to permit reasonably normal gene function in a sharply modified genetic context. In the context of crop plants and weed races or related wild species the nature of this differentiation-hybridization cycle can be important in determining the extent of gene flow and introgression between them, although it is never sufficient to cause disintegration of the two entities. In species which are weakly genetically buffered, such as a self-fertilizing diploid, of which barley (*Hordeum vulgare*) is an example, the normal reproductive process rarely creates new genotypes that differ markedly from those of the parent. The results of crosses between different entities are therefore likely to be major.

Maize, a cross-fertilizing diploid, has more genetic buffering than barley. Crosses occur between parents which have at least slight differences in genetic constitution. For the offspring to survive, their genetic system must have some buffering mechanism that permits reasonably normal gene function in a new context. Consequently, because of the amount of heterozygosity, hybridization can occur in maize between strains that have diverged to a greater degree. Other species such as wheat, sugarcane and the potato, are still more strongly buffered mainly because they are polyploid. The polyploid condition means that they possess more than two sets of chromosomes, and that the genetic constitution has a major element of redundancy in it. This redundancy tends to strengthen the degree of genetic buffering, and such species can withstand massive doses of alien germplasm and rather wide crosses, which interspecific hybridization represents. Such wide crosses are effective in increasing genetic variance, and in vegetatively-propagated crops the loss of sexual fertility may not present a problem.

What is found therefore is a situation of polyploid complexes related to the distribution of so-called diploid pillar species, which are important in terms of plant genetic resources. Traditional agricultural systems have been important in bringing together species which have subsequently hybridized, and often through polyploidization they have been able to increase variation through infusions of alien germplasm, making these plants of increased value to humans.

It is convenient at this point to illustrate these general ideas with specific crop examples, chosen because they are amongst the world's most important crops (Table 2.3).

Table 2.3 The world's top 30 crops (excluding grass).

Crop	Annual production (millions of metric tons)
Wheat	420
Maize	385
Rice	375
Potato	285
Barley	170
Sweet potato	145
Cassava	100
Soybean	95
Grapes	70
Oats	65
Sorghum	55
Sugarcane	55
Oranges	55
Millets	50
Banana	40
Tomato	40
Sugar beet	35
Rye	35
Apples	35
Coconut	30
Cottonseed oil	30
Peanut	25
Yam	20
Watermelon	20
Cabbage	15
Onion	15
Beans	10
Peas	10
Sunflower seed	10
Mango	10

Axis: 0 50 100 150 200 250 300 350 400 450 — Annual production (millions of metric tons)

Wheat (Triticum aestivum — *Gramineae*)

Wheat is the world's most widely cultivated plant. It is the major crop of the USA and Canada, and is grown widely elsewhere in the world far beyond its centre of origin in South West Asia. It was one of the first plants to be cultivated, if not the first, as indicated from archaeobotanical remains found at several sites from the Fertile Crescent in the Near East dated at over 8000 years ago. When they domesticated wheat, humans laid the foundations of western civilization.

Wheat forms the classical polyploid series, based on x = 7, with diploids, tetraploids and hexaploids, having 14, 28 and 42 chromosomes. At the diploid and tetraploid levels there are both wild and cultivated species, but no hexaploid wild species are known, indicating that breadwheat was formed after domestication of the diploid and tetraploid wheats. The major wheat species are listed in Table 2.4. At least three centres of diversity are known, and by Vavilov each of these was considered as a centre of origin, each for a different group of species. However, Vavilov's interpretation was a simplification of the situation as we now understand it.

The diploid einkorn, *T. monococcum* is found now only in relic cultures, although its greatest diversity lies in Turkey, where wild einkorn, *T.*

Table 2.4 The major wheat species.

	Wheat Species Groups			
	EINKORN 2n = 14 Diploid	EMMER 2n = 28 Tetraploid	TIMOPHEEVII 2n = 28 Tetraploid	VULGARE 2n = 42 Hexaploid
Genome	AA	AABB	AAGG	AABBDD
Wild	*T. boeoticum* (= *T. aegilopoides*, *T. thaoudar*, *T. urartu*) (Balkans, Anatolia, Caucasus)	*T. dicoccoides* Palestine to Transcaucasia and Iran)	*T. armeniacum* (Armenia)	
Cultivated	*T. monococcum* (relic cultures: W. Europe, Anatolia N. Africa, W. Georgia)	*T. dicoccum* (relic cultures: Europe, Ethiopia, C. Asia) *T. durum* (Medit. & E. Europe)	*T. timopheevii* (Georgia) *T. zhukovskyi* (Georgia) AAAAGG	*T. aestivum* (= *T. vulgare*)

boeoticum is also found. The tetraploid emmer wheats are found throughout the Fertile Crescent, from eastern Turkey to western Iran. *T. dicoccoides*, wild emmer is found in this area, so the evidence points to the origin of the emmer wheats in this region. Following domestication of emmer wheat, considerable diversity accumulated at the tetraploid level by mutation and selection, and some of the forms have been accorded specific rank. Vavilov found immense diversity of tetraploid wheats in Ethiopia, and considered that this was a centre of origin. However no wild wheats occur there, and cultivated forms must have been carried there by nomadic peoples, after which diversity accumulated.

The greatest diversity of the hexaploid wheats lies further east compared with the diploids and tetraploids, in an area of the western Himalayas, round the Hindu Kush. As no wild hexaploid species occur, breadwheat must have originated by hybridization between cultivated and wild species, after which it spread from its centre of origin, accumulating a large amount of diversity in the process.

Wheat provides an excellent example of the role of hybridization and polyploidy in crop evolution. It is clear that cultivated wheat evolved by selection for characters such as non-brittle rachis from wild species at the same ploidy level, and that the different polyploids were formed following hybridization between wheat species and related species of goat-grass. It is convenient to refer to these species as belonging to the genus *Aegilops*, although Feldman (1976) has assigned them to *Triticum*. It is clear that there are close biological affinities between *Triticum* and *Aegilops*, but for the sake of clarity, *Aegilops* will be retained in this text.

Using the techniques of cytogenetics and electrophoresis of seed proteins, the evolutionary relationships of wheat have, by and large, been determined.

The diploids have been assigned the A genome, the tetraploids the AB genome, and the hexaploids the ABD genome. Whereas the identity of the A and D genomes is known with confidence, with *T. boeoticum* and *Aegilops squarrosa* being the species involved respectively, the identity of the B genome species remains a source of controversy. Sarkar and Stebbins (1956) postulated *Ae. speltoides* as the donor, and compared experimental hybrids with *T. dicoccum*, with which they found reasonable morphological correspondence. On electrophoretic evidence, Johnson (1975) stated that another wild diploid wheat, *T. urartu*, was the B genome donor, and the results of other cytogenetic studies have suggested several different species of *Aegilops*. The identity of the B genome is, however, far from certain. The generally accepted cytogenetic scheme of wheat evolution is shown in Fig. 2.3

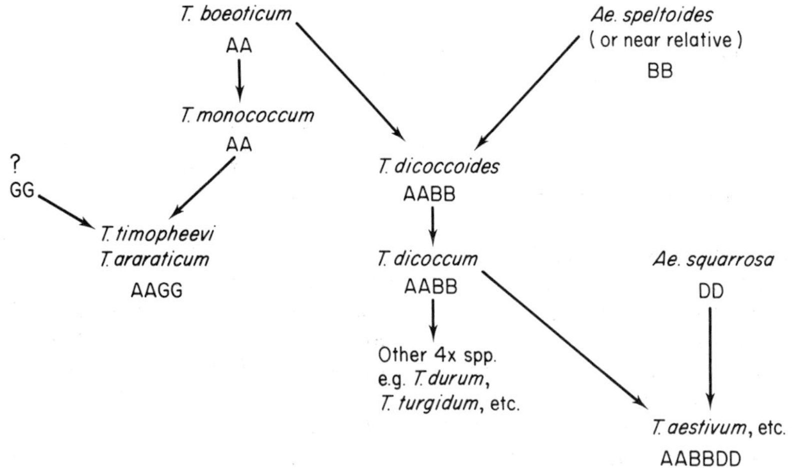

Fig. 2.3 Generally accepted cytogenetic scheme of wheat evolution.

The diploid-like behaviour of polyploid wheats is due to the suppression of pairing of homoeologous chromosomes from the different genomes. A specific gene on the long arm of chromosome 5B in *T. aestivum* is responsible for this phenomenon (Riley, 1965). The importance of this diploidizing mechanism has been of the greatest evolutionary importance in wheat. By restricting pairing to homologous chromosomes, regular segregation of genetic material, high fertility and genetic stability are ensured. The combination of three genomes in a single individual provides *T. aestivum* with greater ecological amplitude than that found in the diploid and tetraploid cultigens, which led to its pre-eminence amongst the cereals.

Maize (Zea mays — *Gramineae*)

Maize was originally the important carbohydrate source of the New World tropics, although it has now assumed an important agricultural role elsewhere in the subtropics and even temperate areas. Although genetic analysis has proceeded more with this species than perhaps with any other species, the story of the origin, evolution and domestication of maize has attracted more controversy than most crops.

Maize is a monoecious species with separate male and female inflorescences on the same plant. It belongs to the tribe Maydeae, and with two other genera, *Euchlaena* (now sometimes included in *Zea*) and *Tripsacum* it is New World in origin. The archaeological evidence for maize is most impressive, with primitive remains dated at 5600 BP from Bat Cave in New Mexico, and other remains from Tehuacan in Central Mexico, at 7400 BP. Do these remains represent wild maize or a domesticated form? The American geneticist Paul Mangelsdorf has tried to explain the origin of maize in terms of a complex of hybridizations involving *Zea mays*, teosinte (*Euchlaena mexicana*) and *Tripsacum* (Mangelsdorf, 1965). On the basis of extant races of maize, he suggested that maize developed from a pod-popcorn, but since domestication, maize has become completely dependent upon man for its survival. It has lost its seed dispersal mechanism completely, and the seeds are enclosed and protected by modified leaf sheaths. While there is little disagreement that introgression from teosinte has occurred (both maize and teosinte have 20 chromosomes), resulting in several morphological characteristics including induration or hardening of the cob rachis, to allow larger heavier inflorescences to develop, the evidence for hybridization with *Tripsacum* is less certain. *Tripsacum* has $2n = 36$. Introgression with teosinte, however, can still be seen at the present time, and teosinte can be found in several places in Mexico and Guatemala as a weed of maize fields.

More recently De Wet *et al.*, (1971), Wilkes (1979) and Beadle (1980) have indicated that the direct ancestor of maize was teosinte. Biologically, maize and teosinte belong to the same species, and some workers have found it convenient to classify teosinte in the genus *Zea*. In crosses between some races of maize and teosinte, parental types have been recovered at a frequency that indicates that the differences between them are the result of as few as five genes. In fact, a careful morphological analysis of the necessary changes for the derivation of polystichous maize from distichous teosinte suggests that these changes are not complex. Teosinte is a wild plant which retains a seed dispersal mechanism. Seeds are protected in hard cupules, and these are arranged in two rows on either side of the rachis. There have been reports that in some areas of Guatemala the seeds of teosinte are heated and utilized much as a 'popcorn', which would suggest a way in which this species was first brought into domestication.

The recent discovery of a diploid perennial species of teosinte, *Euchlaena* or *Zea diploperennis* (Iltis *et al.*, 1979) has further clarified the picture, and has established the basis for the evolution of the annual wild form. Until this discovery the only perennial form known, *Euchlaena perennis*, was an autotetraploid with 40 chromosomes, and was therefore not considered as a progenitor of maize.

Following domestication there was differentiation into a number of distinct races, such as pod corns, popcorns, flint, dent and sweet corns. Considerable diversity accumulated in Peru, although there is no evidence to suggest that maize originated there. The Peruvian maizes are thought to be relatively 'pure' showing none of the so-called 'tripsacoid' features of their Meso-American counterparts, as a result of introgression with teosinte.

Despite recent discoveries, the controversy over the origin of maize still persists. What is certain is that this important crop has had a long history of utilization in the Americas, upon which the development of several important cultures, particularly the Aztec culture, depended.

Rice (Oryza sativa — *Gramineae*)

Rice is perhaps the most important crop grown in the world at the present time. More people depend upon it directly as a food than any other crop. It is the staple crop in S.E. Asia, in China and parts of the Indian subcontinent, where world population is growing fastest.

There are two species of rice, Asian rice *O. sativa*, and African rice *O. glaberrima*, and although obviously closely related on the basis of cytogenetic studies, hybrids between the two species are sterile. Both are diploid, with 14 chromosomes, and the tetraploid species of *Oryza* have played no role in the evolution of the cultigens. The genus is world-wide in distribution, but species in the Americas and Australia can be discounted in rice evolution.

The pattern of evolution is remarkably similar and parallel in both Asian and African rices. Both have involved a sequence of wild perennial forms giving rise to annual forms, from which domesticates have been selected. In the case of African rice, this sequence is *O. longistaminata* and *O. barthii* giving rise to *O. glaberrima*. African rice has achieved little more than local importance, and has recently been replaced to a considerable extent by Asian rice. Asian rice was derived from *O. rufipogon* and *O. nivara*, and as with *O. glaberrima* there are weedy races with which gene exchange does occasionally occur. In the case of Asian rice there has been a further differentiation into three important eco-geographical races, *indica*, *javanica*, and *japonica* or *sinica* (Chang, 1977). This differentiation is manifested by various characteristics such as photoperiod requirement and morphology, and reproductive barriers have evolved to isolate these races.

The time-scale of rice domestication is much harder to define than for cereals such as wheat and maize which evolved in drier climates where organic remains have been preserved. Much of the evidence for rice domestication is circumstantial, based on artifacts which have been found. The evidence for Asia is better than for West Africa, where little archaeological research has been undertaken relative to other parts of the world, even though it is known that several important cultures developed in that part of the world.

Barley (Hordeum vulgare — *Gramineae*)

Barley is an important grain crop, grown in temperate parts of the world mainly as an animal feed, malt for beverage and food products, and human food. It grows well where the ripening season is cool, rainfall moderate, and soil medium-well to well-drained. Major producing countries are the USSR, China, Canada, France, United Kingdom and USA. Spring barley is the predominant form grown, with winter types being grown in regions with mild winters.

The grain-producing forms fall into the section *Cerealia* of the genus. Because of virtually complete fertility occurring between all the different forms within this section the wild and domesticated forms probably represent various genotypic combinations of one single polymorphic species (Harlan, 1979). These are therefore classified as one species, namely *Hordeum vulgare* (Bowden, 1959). This encompasses the earlier recognized wild species, namely *H. spontaneum* and *H. agriocrithon* and the cultivated types, *H. vulgare* and *H. distichon*.

The earliest positive archaeological remains were of wild barley (*spontaneum*) found at Ali Kosh in Iran and dating back as early as 9500 BP. Two-rowed cultivated forms were also identified and dated to about 7000 BP. However, carbonized spikelets similar to the '*spontaneum*' form have been recovered from Tell Murreybit in Syria and possibly can be dated to 10 000 BP.

The first speculation as to the origin and evolution of cultivated barleys was by De Candolle (1882) who proposed that six-rowed cultivars would have arisen either from '*spontaneum*' forms in very ancient times, or from an extinct wild 6-rowed progenitor. These two alternative lines of origin have since been the basis of much controversy.

Schiemann (1932) has postulated on morphological grounds that a then-undiscovered 6-rowed wild barley might have given rise to 6-rowed cultivars as well as the wild '*spontaneum*' form. The subsequent discovery by Åberg in 1938 of *Hordeum agriocrithon*, a wild 6-rowed barley, could have lent support to Schiemann's views, but in fact fueled further debate. Åberg (1940) proposed two further hypotheses taking into account the '*agriocrithon*' discovery. His 'monophyletic' hypothesis suggested that '*agriocrithon*' may have given rise to '*spontaneum*' on the one hand and '*vulgare*' on the other, and from which the 2-rowed '*distichon*' developed later. Alternatively, his 'diphyletic' argument proposed that two-rowed cultivated barleys arose separately from *H. spontaneum*.

Zohary (1959) and more recently others (Harlan and Zohary, 1966; Harlan, 1971b, 1979) take a simplistic view of the domestication of the crop which may be summarized by saying that it originated in a limited geographical area (South West Asia) from a 2-rowed ancestor, and that secondarily, and contrary to earlier views, 6-rowed 'wild' barleys such as the '*agriocrithon*' type reflected hybridization and introgression with cultivars.

Although the idea of a 2-rowed ancestor is likely to be correct, recent evidence suggests that both present-day '*spontaneum*' and '*agriocrithon*' types have diverged from the same ancestral stock through gene mutation, migration and drift. This is suggested to be the case by Shao (1981) who studied barleys in the Tibet-Himalaya region and found '*spontaneum*' and '*agriocrithon*' growing sympatrically, but reproductively isolated. It is further supported by Ayad (1983) who has made estimates of genetic distance and gene diversity of hordein loci in a wide range of wild and cultivated material, and who disputes the hybrid origin of '*agriocrithon*'. The process of gene mutation coupled with migration and drift, rather than inter-group hybridization, is considered as most likely to have been the evolutionary mechanism responsible for the divergence of barley landraces or wild forms.

Potatoes (Solanum tuberosum — *Solanaceae)*

The potato is the most important root crop grown in the world. In South America, it represents the staple food of the Quechua and Aymara Indians of Peru and Bolivia, and it was their ancestors who domesticated the potato. Although potatoes are found throughout the Andes from Venezuela to Chile and north west Argentina, it is from central Peru to central Bolivia, and particularly in the area of the Lake Titicaca Basin that the greatest diversity of cultivated forms is found. The common potato, *Solanum tuberosum* (a tetraploid species with 48 chromosomes) is represented by two subspecies *andigena* and *tuberosum*, the former being indigenous to the Andes, and adapted to short days. The latter is adapted to long days and is found in Chile, and also developed in Europe after its introduction in the sixteenth century, although there is no evidence to suggest that the potato was taken first from Chile to Europe.

There are also four diploid cultigens, *S. phureja*, *S. stenotomum*, *S. goniocalyx* and *S.* × *ajanhuiri*, two triploids, *S.* × *chaucha* and *S.* × *juzepczukii*, and one pentaploid species *S.* × *curtilobum*. In addition, there are almost 200 wild species of potatoes which are distributed throughout the Americas, from the south west United States, through Mexico and Central America, and down the Andes into north west Argentina and even into the plains of Paraguay, Uruguay and southern Brasil.

Potatoes were cultivated at least 8000 years ago, as archaeological evidence from sites in Peru has indicated. Martins (1976) identified material from the Chilca canyon, and material from coastal sites suggested contact between coastal communities and agricultural cultures in the high Andes. Pre-Columbian ceramics, retrieved from graves which are common along the coast of Peru also indicate the esteem in which the potato was held, as ceramic representations of this and other crops are frequently found (Fig. 2.4).

But what of the origin of the cultigens? Comparative studies of the diploids suggest that *S. stenotomum* is the most primitive species, and Hawkes (1978) has suggested that *S. phureja* and *S. goniocalyx* were derived from it by mutation and selection. Indeed there are no genetic barriers separating these species. *S.* × *chaucha* is apparently a series of triploid genotypes formed by hybridization between tetraploid and diploid cultigens (Jackson *et al.*, 1977). *S.* × *ajanhuiri*, *S.* × *juzepczukii* and *S.* × *curtilobum* are all bitter-tasting, frost tolerant potatoes which must be processed before consumption. Huaman *et al.*, (1980, 1982) have shown that *S.* × *ajanhuiri* is a hybrid between *S. stenotomum* and a wild species *S. megistacrolobum*, which was formed in the Lake Titicaca region. *S.* × *juzepczukii* and *S.* × *curtilobum* derive their frost tolerance from a tetraploid wild species, *S. acaule*. *S.* × *juzepczukii* is a hybrid between *S. acaule* and *S. stenotomum*, whereas *S.* × *curtilobum* is probably derived from crosses between *S. tuberosum* subsp. *andigena* and *S.* × *juzepczukii*, in which an unreduced gamete functioned (Hawkes, 1962; Schmiediche *et al.*, 1982). The relatively small number of clones of these hybridogenic species does suggest that hybridizations of this nature have not occurred very often.

Fig. 2.4 A Chimu culture ceramic pot(*c.* 14th century AD) depicting potato tubers, from the northern coast of Peru. Much earlier ceramics in the form of potatoes and other food crops have been found at several other sites in Peru.

S. stenotomum was probably derived from a wild diploid species, such as *S. canasense* or *S. leptophyes* (Hawkes, 1978), and through a process of hybridization with a related weed race, *S. sparsipilum*, and subsequent selection, the tetraploid cultigen was formed (Cribb, 1972). The species relationships are summarized in Fig. 2.5.

Potato species have not evolved the degree of genomic differentiation which is found in many other crops. Instead potato species are eco-geographically isolated, but when they do come into contact the potential for hybridization does exist. In addition, it is now clear that sexual polyploidization has played a major role in potato evolution, and that unreduced gametes, which have the somatic chromosome number, are a common phenomenon in many species (Quinn *et al.*, 1974). As a result, wild potato species represent a major genetic

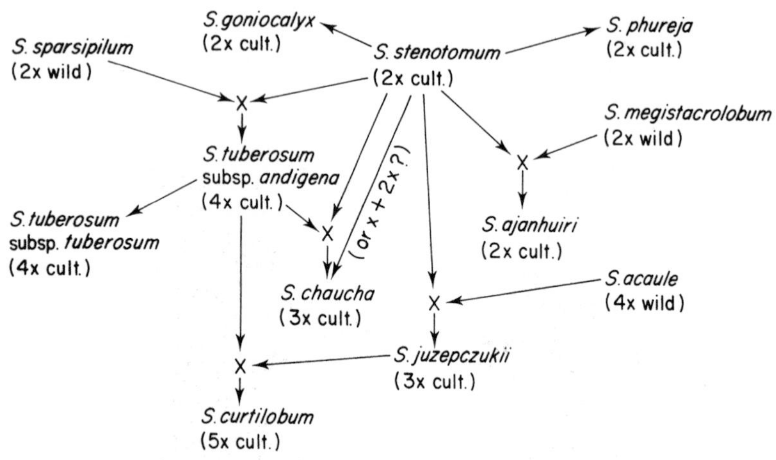

Fig. 2.5 Evolutionary relationships of cultivated potatoes.

resource, and as we shall see in Chapter 6, potatoes are one of the main crops in which the exploitation of wild germplasm in breeding has been important.

Cassava (Manihot esculenta — *Euphorbiaceae*)

Cassava is an important tropical crop plant, accounting for 59% of tropical root and tuber production (FAO, 1976), and which produces high yields of carbohydrates from its tuberous roots. The genus is found in the Americas, and the cultigen *Manihot esculenta* has two important centres of diversity in south and east Brazil and in southern Mexico. Although a crop of American origin, it has now achieved importance as a root crop throughout the tropics.

Cassava is extremely varied morphologically, and appears to be a segmental allotetraploid with 36 chromosomes. Many forms of cassava are toxic, and this toxicity is associated with free hydrogen cyanide (HCN) after cyanogenic glycosides have been hydrolysed. However, the HCN content does not correlate with morphology. So-called sweet varieties are known, but bitter varieties can be eaten if processed.

It has been suggested that the ancestors of cassava were among the first plants to be used as food when man migrated into Central and South America (Jennings, 1976). Comprehensive taxonomic studies of *Manihot* by Rogers and Appan (1973) indicate that three species, namely *M. aesculifolia*, *M. rubricaulis* and *M. pringlei* are the closest relatives of cassava, and these have several characters in common with the cultigen, including in the case of *M. pringlei*, a low content of cyanogenic glycosides. It is apparent that hybridization between the cultigen and wild species has amplified the variability of the former. Wild species have weedy tendencies, and by invading disturbed areas, such as those used by humans for cultivation, the chances for hybridization would be increased. There appear to be few barriers to hybridization between species of *Manihot*, except where the cultigen itself is sterile, so that the potential for gene exchange is great, and the monoecious habit of the plants and protogyny would enhance the chances of natural hybridization.

Jennings (1976) has listed the changes which occurred under domestication, including selection for large roots, erect and less branched growth, and the ability to propagate from stem cuttings. Because of the reduction in branching, flowering was reduced, and many forms virtually became sterile. Although there was selection for a reduction of glycosides, no forms are completely free of the poison, and it is clear that in some areas, the bitter varieties are actually favoured. It has been postulated that the bitter varieties were first used as a fish poison, and only later as a food.

Archaeological remains of seeds and leaves dated at about 2100 BP from north eastern Mexico have been identified as cassava. However, remains from the Peruvian coast dated at 4500 BP were identified by Martins (1976) as cassava by comparing starch grains from fresh and archaeological material. Because of the widespread nature of the cultigen, and the pattern and concentration of diversity displayed in at least two areas, there is considerable support for a dual if not multiple origin of this important root crop.

Sweet potato (Ipomoea batatas — *Convolvulaceae*)

The sweet potato belongs to a large tropical and sub-tropical genus, *Ipomoea*. It is a plant which bears root tubers, and is propagated vegetatively through the root tubers, or more commonly by stem cuttings.

There is general agreement that the sweet potato is of American origin, and its closest relatives such as *I. triloba*, *I. trifida* and *I. tiliacea* are American. The section *Batatas* which includes the cultigen, has a centre of diversity in Mexico. Cytologically, *I. batatas* is a hexaploid with 90 chromosomes, as is *I. trifida*. The cytotaxonomic relationships of the species are still the subject of controversy, and the evolutionary steps by which the crop originated remain uncertain. Undoubtedly the polyploid species were formed following hybridization between lower ploidy species. If *I. trifida* is the ancestral form of *I. batatas*, or if both species evolved from some ancestral stock, the origin of these hexaploids still needs to be discovered.

Sweet potato does provide an interesting example of crop dissemination. Archaeological remains of sweet potato tubers have been found in Hawaii, New Zealand and Easter Island, although the date of the spread of the crop throughout Polynesia is still not certain. Yen (1974, 1976) has suggested that it spread before the eighth century A.D. but a post-Columbian spread is favoured by other workers. What is certain is that in New Zealand the crop was converted into an annual plant in response to seasonal temperate climates in which tropical cultivation practices were impossible to apply. The voyages of the Polynesians and the voyage of the Kon-Tiki indicate that dissemination of the sweet potato could have occurred before the early Spanish voyages after the discovery of the Americas.

To a certain extent the point of origin of the sweet potato remains an enigma. The development of forms adapted to regions with marked wet and dry seasons which would produce swollen roots as food reserves, suggests the area of the summer-green forest and scrub belt which spreads round the Caribbean from Mexico to Venezuela and north Brazil. In fact the word 'batatas' is a Caribbean Arawak Indian word, and the introduction of the sweet potato into Europe was from that region. So although we cannot point to one area with certainty, Mexico is a likely candidate because of the presence there of *I. trifida*.

Grain legumes or pulses (Leguminosae)

A pulse is the dried edible seed of a cultivated legume. Pulse crops are of very ancient origin in both the Old World and the New World. Their agricultural importance lies in their protein content, and they contain more protein than any other plant product. Animal protein is still a rarity in the diet of vast numbers of poorer people in the tropics, and pulses provide the chief, if not the only source of protein. Most pulses are harvested when the seeds are ripe and the pods are dry. Their low moisture content and hard testa permit storage over long periods.

The Vicieae and Phaseoleae are the main taxonomic groups which have contributed pulses to agriculture, and all the important pulses belong to these two tribes (shown in Table 2.5), except the groundnut, *Arachis hypogaea*, which belongs to the Hedysareae.

Table 2.5 Some food legumes important in world agriculture.

Tribe Vicieae	
Pisum sativum	Field and garden pea
Vicia faba	Broad bean
Lens culinaris	Lentil
Cicer arietinum	Chickpea
Lathyrus sativus	Grasspea or Khesari dahl
Tribe Phaseoleae	
Phaseolus vulgaris	Common bean
P. coccineus	Runner bean
P. lunatus	Lima bean
P. acutifolius	Tepary bean
Vigna unguiculata	Cowpea
Cajanus cajan	Pigeon pea
Glycine max	Soybean
Psophocarpus tetragonolobus	Goa or winged bean
Tribe Hedysareae	
Arachis hypogaea	Groundnut

It is interesting that the evolutionary trends shown by the pulses in the course of their domestication are closely parallel, and according to Smartt (1978) present a good example of Vavilov's Law of Homologous Series. Characteristic features of individual pulses not found in their closest wild relatives can almost invariably be found in other members of the same tribe, or even other tribes. A common pattern of evolution under domestication is clear. A number of evolutionary trends are apparent which affect the vegetative parts, including changes from diffuse to compact growth habit, erect growth habit, and gigantism of these vegetative parts. Also of great importance are the changes which have altered the reproductive strategies and structures of the pulses. These include changes from a perennial to annual life form, increases in seed and pod sizes, reduction in dehiscence of pods, loss of seed dormancy and loss of photoperiodic sensitivity.

Not all of these factors have affected all pulses, but in species of *Phaseolus*, the differences between wild and cultivated forms are considerable. In several pulses, the closest wild relatives and progenitors are well known. In *Phaseolus*, for instance, there are conspecific wild forms of three of the cultigens, whereas in species such as *Vicia faba* and *Lathyrus sativus* the wild progenitor is not known with any certainty whatsoever. In other genera such as *Pisum* and *Cicer*, the wild progenitors have been identified on the basis of detailed cytogenetic and experimental hybridization studies (Ben Ze'ev and Zohary, 1973; Ladizinsky and Adler, 1976).

While it is possible to make general statements about evolutionary trends in the grain legumes, detailed information concerning the evolution of individual crops may not be available. The clearest and most complete picture based on data from archaeology, palaeoethnobotany, botany and genetics is provided by the species of *Phaseolus* beans, in which there are forms which have contributed cultigens to agriculture and others which have not. In their mode of life and growth habit the wild species resemble each other, and often they are found growing as climbers amongst shrubby vegetation.

Phaseolus beans are ancient cultigens in both Meso-America and South America, and the archaeological record is most impressive, with the oldest remains dated at over 7000 years old. At the present time, *P. vulgaris* is the most important cultigen in the genus, and it now has an importance far beyond the Americas. In this species, its success relative to the three other cultigens can be attributed to several factors. The initial gene pool of the species is larger than the other species because of the widely distributed wild forms. The present extensive gene pool of *P. vulgaris* owes something to the wide distribution of its progenitor and the possibility that domestication has occurred repeatedly over much of its geographical range. The range of variation manifested in *P. vulgaris* is considerable in growth habit, seed shape, seed coat colour and pod texture. The pre-eminent position of this species may be due to an inherent capacity to respond to selection, coupled with the heavy selection pressures applied. Its rapid spread in post-Columbian times may be a result of its wide dispersion and selection over a range of temperate environments in earlier pre-Columbian times (Smartt, 1969).

Conclusions

The preceding discussion has illustrated the array of evolutionary processes which have played an important role since crops were first domesticated. In many areas of the world where traditional varieties can still be found in contact with related wild species, hybridization and introgression continually lead to gene-flow between species. This dynamic system continues until modern agricultural practices and varieties replace traditional ones, and 10 000 years of natural evolutionary change are brought to a conclusion. The importance of plant genetic conservation can easily be appreciated in this context!

3

Plant Genetic Resources Exploration

Early days — the botanical collectors

The earliest records of plant collecting expeditions date back to 1495 BC when Queen Hatsheput of Egypt sent an expedition to Somalia in search of trees whose fragrant resin yielded the precious frankincense (Coats 1969). Other records chronicle the activities of various collectors during the Dark Ages, and during the 17th century the collections made by the Tradescants, father and son. The golden age of plant collectors however was the eighteenth century, when botanists under the direction and patronage of Sir Joseph Banks (1743–1820), President of the Royal Society, and of the royal garden at Kew (now the Royal Botanic Gardens) were sent to various parts of the world in search of exotic species. Banks himself had sailed with Captain Cook in 1768 to the Pacific islands, and in later years he instilled into hopeful plant collectors the dedication required for plant hunting, the single-mindedness, the stamina and the cheerful indifference to discomfort and to continuous disappointment. In those days, plant collecting was a dangerous occupation with little guarantee that the plants collected would be safely returned to England.

Among the collectors who prospered under Banks' patronage, three are worthy of special mention, because they exemplify the difficulties faced by such men. Francis Masson (1741–1806) collected twice in South Africa, Madeira, the Canaries, the Azores and West Indies, and met his death in North America. He was responsible for the introduction of Pelargoniums into England from South Africa, as well as varieties of Proteas and Aloes.

David Nelson (d. 1789) accompanied Captain Bligh on the fated voyage of HMS Bounty to the Society Islands, to collect breadfruit for planting in the West Indies, as a cheap and palatable food for the slaves. Perhaps the breadfruit was the indirect cause of the mutiny which followed. The propagation of the plants ensured a long stay in Tahiti, after which the crew were reluctant to leave. Nelson was cast adrift with Bligh in a small boat, and although Bligh managed successfully to navigate 4500 miles to safety, Nelson died of lung fever before the end of the voyage.

George Caley (1770–1829) hunted alone for plants during a period of ten years in the remotest parts of Australia, and he sent many valuable seeds and stocks to Kew, together with detailed botanical descriptions. The introduction of exotic species at this time was paralleled by advances in systematic botany. After the death of Banks in 1820, collecting activities were reduced and the financial allocation to Kew was cut. This state of affairs continued until 1841 when Sir William Hooker became the first official Director of the National Botanic Garden.

Parallel collecting activities were proceeding apace in other parts of Europe, with the establishment of botanical gardens and the expansion of colonial administrations overseas. The nineteenth and early twentieth centuries also saw the work of collectors such as George Forrest (1873–1932) and Peter Barr (1825–1909). Forrest made various expeditions to the Himalayas and China, and brought back many species of *Primula* and Rhododendrons. For Barr, the ultimate love of his life was the *Narcissus*, and he set himself the task of rediscovering all the species mentioned by John Parkinson in his *Paradise in Sole*, published in 1629, many of which had been lost to cultivation for more than 200 years. His travels took him to many areas where wild daffodils might be found, in Spain, Portugal and the Maritime Alps.

As a result of the activities of these and other plant hunters, many exotic species were introduced into Europe which are now widely known in cultivation. This explorative phase of different floras provided the basis for a sound systematic botany. But although these collectors were concerned with diversity, it was diversity in the sense of collecting as many different species as possible. Post-Darwinian appreciation of the variable nature of plant species meant that botanists could no longer take single plants as being representative of a species, although for taxonomic purposes, this is often what a plant collector may do even now. Nevertheless today, many botanic gardens play an important role in *ex situ* conservation of endangered species (see Chapter 4).

Early germplasm collectors

It was Vavilov and his co-workers who realized the value of diversity in crop plants and their wild relatives for plant breeding purposes, and during the 1920s and 1930s expeditions were sent to many parts of the globe in search of crop diversity. From a study of the material collected, Vavilov was able to formulate his hypothesis on the 'centres of origin' of crop plants. Following in Vavilov's footsteps have been other famous collectors such as Harry V. Harlan (barley) and his son Jack R. Harlan, Hermann Kuckuck (wheats) and Jack Hawkes (potatoes) amongst others. The enthusiasm for plant exploration was described by H.V. Harlan (1957), who wrote in his book *One Man's Life with Barley*, 'The desire to explore and to collect is as inherent in us and as much part of us as toenails and far less transitory than hair. Who has not felt the urge to go places and do things? I wanted to go so badly that I would not have blamed those in power if they had harbored a suspicion that my eagerness was as much for personal pleasure as for the good of the cause'. Certainly this enthusiasm has been shared by many collectors since Harlan, and has provided the basis for the successful collection of plant genetic resources often under circumstances of personal discomfort and danger.

Genetic resources exploration

It is now realized that the exploration and collection of plant genetic resources must be based on the application of sound scientific principles. The objectives of the botanical collector and the plant genetic resources collector are not the same. Whereas the botanical collector may look for uniformity or trueness to type, the field sampling procedures in plant genetic resources exploration are aimed at the fullest possible recovery of genetic variation of species, irrespective of the relative frequency or rarity of any genes or linked genetic complexes. Marshall and Brown (1975) have indicated that an optimum sampling strategy should enable the collector to obtain, with 95% certainty, all the alleles occurring in the population at a frequency greater than 5 per cent. In order to achieve this, it is desirable to understand something of the population structure of a species and its eco-geographical distribution. While considerable research has been undertaken with annual seed crops, especially grasses and cereals, it is unlikely that the conclusions drawn from these can be applied to the situation which obtains in vegetatively-propagated crops, such as potatoes and yams, in fruits and tree species and in forages. Conservation strategies for all of these differ in certain aspects, and consequently affect the exploration strategy. Nevertheless there is a number of general exploration premises which apply to sampling all types of crops.

Sampling strategies

It can be argued that the theory of sampling strategies must be founded upon an extensive knowledge of the patterns of genetic variability of populations. However, there are relatively few species for which this type of information is available. What can be said, and what has been found for most species is that they exhibit extensive geographical variation and that superimposed on this is a pattern of extensive variation within populations. Ecological factors are a major determinant of genetic diversity, and many species display differentiation into ecotypes in response to their habitats (Gregor, 1933). This is also characteristic of the agricultural environment, and agro-ecotypes are most clearly distinguished in primitive cultivars. Climatic factors such as maximum and minimum average temperatures, precipitation and the seasons of dormancy and growth, light intensity, and day length are all reflected in corresponding developmental characteristics (Bennett, 1970). Such factors generally lead to clinal variation patterns, whereas topographically or edaphically determined differentiation may lead to either a clinal (as in the case of altitude) or a mosaic distribution. Consequently the distribution of variation as affected by such factors should influence the frequency of sampling.

Any collection can only be a small sample of the total variation. Geographical variation patterns include such characteristics as disease resistance, morphological features and other conspicuous differences as well as variation in quantitative characters. The relevance of these to plant breeding is obvious, for although varieties or strains may look alike, they may differ greatly in factors useful to the plant breeder, especially in differences in physiological characters. The variation within populations, especially at the local level, will be controlled by the interaction of the breeding system of the species and the forces by which variation is maintained. The maintenance of heterozygosity in outbreeding species is clear. Populations of inbreeding species are commonly supposed to have a simple genetic structure consisting of a number of inbred lines each of which maintains itself as a constant, genetically homozygous entity for a number of generations. But as Allard (1970) has indicated, there is now a considerable body of evidence which shows that such populations contain much genetic variation and heterozygosity. Sampling methods must be used to ensure the collection of representative within-population variation as well as that associated with geographical patterns of variation.

A number of studies have been undertaken on the nature and distribution of variation in wild species populations, and from these it is possible to formulate recommendations for the overall sampling strategy. First we shall deal with the questions relating to the 'nuts-and-bolts' of sampling, which involve the number of plants to sample per site, the total number of sites to sample and the distribution of sampling sites within each area. It is also convenient to discuss the different types of crops separately.

Seed crops

A rational assessment of the sampling strategies for seed crops must take into consideration all available resources in terms of personnel, time and finance. Various authors, including Bennett (1970), Allard (1970), Marshall and Brown (1975) and Hawkes (1980) have given precise details of the number of plants to sample at each site, and Marshall and Brown have even stated that the relatively large samples recommended by Bennett and Allard (200–300 plants per population) represent extreme wastefulness. They indicate that, in most circumstances, sample size should not exceed 50 plants per population and in no circumstances is it desirable to collect more than 100 plants per population. We cannot altogether agree with their point of view, and although collecting in the field becomes somewhat of a compromise between resources, we recommend that the sample size should be as large as possible. In situations where populations are not so large as they are in the cereals, it may not be possible to collect a large sample, or for instance the farmer may not be willing to give more than a small sample from his seed store. We feel that the recommendation of arbitrary sample sizes promotes confusion in the mind perhaps of the less experienced collector, and consequently he or she may not do a good job in the field. The collection of samples which are too small carries with it the great problem of genetic drift upon consequent immediate regeneration of the seed. Larger samples enable at least a proportion to be put into long-term storage right away, avoiding the need for immediate regeneration.

Should random or selective sampling be used? Actually random sampling is better referred to as **non-selective** sampling. Truly random sampling as a strategy may only be applicable to wild vegetation, as used in ecological studies, but could not be utilized in a field of primitive wheat, for instance, without causing much damage. This would surely be unacceptable to a farmer! How is non-selective sampling actually carried out? Again most recommendations have been made for cereals. The collector enters the field and works his or her way back and forth across the field until the whole area has been sampled. Samples of seeds are collected every few paces, depending on the size of the sample to be collected, and placed together in a bag.

Without any prior knowledge about population structure, one can recommend that only non-selective sampling will allow the collector to capture most of the common alleles in the population, and this is the primary objective of collecting genetic resources. In addition, however, we recommend the collection of a **selective** sample, whereby phenotypes which are distinguishable, whether in terms of a morphological character, or the manifestation of a physiological character, such as drought tolerance or disease resistance are collected separately. There is some disagreement among germplasm collectors as to whether disease resistant plants can actually be selected in the field, as the definition of resistance can be made only after several observations under experimental conditions. While we accept that this is the case, we do urge germplasm collectors to use their eyes, and if a particular plant looks interesting because of some phenotypically expressed characteristic, then it should be collected as a **selective sample** and maintained separately, in case it does have some value for plant breeding. Often such plants will carry the rare alleles. Marshall and Brown (1975) have indicated that only a bulk sample should be collected from each population. While this is acceptable for the random or non-selective sample, the selective samples should be retained as separate collections, otherwise there is the risk of quickly losing such rare alleles with little hope of retrieving them at a later date for breeding purposes. Marshall and Brown have also stated that the aim of genetic conservation is to collect and conserve only adaptive gene complexes. We feel the conservation of the rare alleles is equally important because it is precisely these alleles which confer resistance to disease for example, which the plant breeder may want to utilize. Furthermore, adaptive complexes are likely to be broken up during the breeding process anyway. Perhaps the conservation of adaptive complexes is particularly important only in crops such as the forage grasses, where less breeding has been undertaken, and selection from plants collected in the wild is the principal means of forage improvement (see Chapter 7).

Vegetatively-propagated crops

With the vegetatively-propagated crops, rhizomes, corms, bulbs, tubers and roots are collected. Why bother with such material? Often such crops are not sexually fertile and do not produce seed; consequently vegetative propagules are the only plant parts available. Even with crops such as yams (*Dioscorea* spp.) which generally have low fertility or which do not flower frequently, it has been possible to devise techniques to promote flowering, as work at IITA has shown. In this way variation is released and this valuable germplasm can be exploited by plant breeders.

It cannot be assumed that the population structure of a vegetatively-propagated crop conforms to the same pattern as a seed crop. Each plant provides only a single type, and as has been shown in the case of potatoes for example (Jackson *et al.*, 1980), a field may contain several genotypes of which one or two comprise the majority of the plants.

Vegetatively-propagated material is bulky and difficult to transport; it is difficult to keep alive, and must be collected at the right stage of maturity or it will not grow. Because these propagules are produced underground, no plant is visible at maturity, and it is easy to see what special problems this causes in the collection of wild material. It is less difficult to collect cultivated material, because of the better defined population structure within a cultivated field.

There is a danger of sampling frequent genotypes more than once and having duplicates, and too many samples in a restricted area will mean that a clone may be picked up several times. This applies also to forage grasses. Notwithstanding these problems, it may be possible to sample all genotypes in a farmer's field or store, and consequently **all** alleles, something which is probably impossible for seed crops. Different genotypes can be recognized due to morphological difference, especially in the part of the plant of economic importance, for example potato tubers. Jackson *et al.*, (1977) have recognized 'morphotypes' in potatoes, and this concept can probably be applied to other crops. The morphotype is conceived as being composed of plants which are similar or apparently phenotypically identical, without implying genetic sameness. In order to be certain of this, more tests would have to be conducted, such as electrophoresis of tuber proteins or disease reaction. It is also our experience that the same morphotype can have different local names, or that the same name is applied to different morphotypes. When presented with this situation, the collector should collect all types which have been identified as different, whether this difference be morphological or linguistic. Duplications can be determined at a later date.

Imagine the problems of transportation of many of these crops. Yams (*Dioscorea* spp.) produce very large, up to 1m, stem tubers. It is not possible to carry such propagules for long distances. Alleviating this transport problem by the removal of the meristematic area, from which sprouts will grow, presents another set of problems relating to disease transmission, especially virus diseases if cutting instruments are not sterilized, but also to contamination and decay of the tuber pieces by other microorganisms. A crop such as cassava is propagated by stem-cuttings, as is sugar-cane, and propagules such as these must be collected. Nevertheless, the application of tissue culture based collecting methods for such species is being investigated. In cocoa, for instance, using a simple method, shoot explants can be kept alive for up to ten weeks (Withers, personal communication). When collecting recalcitrant species such as cocoa, and in very large areas such as the Amazon Basin where communications are difficult, the application of tissue culture based methods has considerable appeal.

If there is some seed production, as is commonly the case with potatoes, for example, then it is advisable to supplement the vegetative material with seeds as well.

Fruit and tree species

Fruit and economic tree species present a special problem. Most temperate and tropical fruit trees, including rosaceous fruits (apples, pears etc.), coffee, tea, cocoa, rubber, most nut trees and most forest trees possess 'recalcitrant' seeds (see Chapter 4) which cannot be stored under normal conditions, and they must be planted soon after collection. The collection strategy must be related directly to the conservation strategy, and sample sizes as recommended by various workers for seed crops cannot be applied to these species. It has been said that the tropical rainforest is an ecosystem of rare species. Species do not occur in dense stands, and a considerable distance may have to be travelled in order to find another individual of a species. Consequently collecting is very difficult under these conditions. Random sampling cannot be applied sensibly when a species is so widespread, but in conditions where individuals are locally frequent, random sampling will improve the capture of alleles in the population. Biased sampling has been used by arboriculturalists and foresters in particular. Collections of seed have been made from the best-looking trees in terms of commercial timber growing characteristics. While such collections satisfy the needs of the forestry industry, such samples are not as highly valuable from the genetic resources point of view. Trees which have poor forestry characteristics may well have valuable disease or pest resistance, and collections should also be made from these (see Chapter 8).

If seeds are not available, or it is undesirable to make collections of these 'recalcitrant' seeds, it is possible to take cuttings of one sort or another from many species. Budwood cuttings, which are lignified but not mature can be collected during the summer months, and then grafted on to rootstocks back at the genebank. The collection and conservation of temperate fruits trees is well documented and strategies have been determined (Sykes, 1975), but the position for forestry genetic resources cannot be said to have developed to the same level. Even less has been worked out for species of the tropical forests.

Forage grasses

With regard to the collection of forage grasses, such as *Festuca* and *Lolium* species, the primary objective has been to sample in localities where populations have evolved under a combination of climatic and other environmental factors. These factors are those considered most likely to have resulted in the combination of characters, i.e. adaptive complexes, required by the breeder (Tyler *et al.*, 1982). Expeditions from the Welsh Plant Breeding Station have collected material from throughout the UK, Belgium, France, Switzerland, Italy, Spain, Norway and Romania (see Chapter 7) in order to obtain material adapted for instance to growth under relatively high winter temperatures, for maximizing growth under high summer temperatures and high light intensities, and from regions with low winter temperatures and relatively low autumnal and winter light conditions which should provide winter-hardy material. While this material has been collected primarily to support breeding programmes, its collection has been of importance to genetic conservation (Chapter 7).

Random sampling is used to collect forage grasses, to ensure that the samples are not too closely spaced because of the spreading nature of grasses by rhizomes, and thus too many duplicates are avoided. The collectors work their way backwards and forwards across a meadow making collections every few metres, the distance between samples being determined on the basis of a quick visual survey before collecting. As with all other types of collecting it is important to use one's eyes continually. Forage grasses may show great ecotypic differences within very small distances. For example, in areas around drinking troughs which cattle frequent, or in well-marked paths, it is possible to find prostrate ecotypes of the more common forage species, and these should be collected, as they may provide valuable genes for resistance to trampling, a character important in forage grass breeding. In areas where cattle congregate ecotypes adapted to high nitrogen levels might also be found.

More crop specific details on germplasm collection are given in several chapters in Frankel and Bennett (1970) and Frankel and Hawkes (1975). In addition a couple of useful pocket manuals on collecting genetic resources have been written by Chang *et al.* (1972) and Hawkes (1980).

Collecting sites

Four main collection sites can be recognized: namely (*i*) farmers' fields; (*ii*) kitchen or orchard gardens; (*iii*) markets; and (*iv*) wild habitats. Most sampling strategies have, in fact, been developed for use in farmers' fields, and this is where the majority of collections of primitive cultivars of the main food crops have been made. Much of what we have said so far relates to the collection of seed crops such as the cereals, which often grow in quite large dense stands and from which large samples can be obtained. Kitchen or orchard gardens, such as that described in the village of Santa Lucia in Guatemala by Edgar Anderson (1969), are particularly important for minor crops, vegetables, horticultural crops, fruits, herbs, condiments and medicinal plants. In areas where genetic erosion has been intense and the widespread cultivation of modern cultivars is common, such gardens can be an important source of genetic diversity. We ourselves have seen, for example, areas in the highlands of Peru where truly primitive potato varieties can only be found in small plots around each farmer's house. The potatoes of commerce, grown on a relatively large scale, are found to be more modern cultivars. Sometimes a collector may arrive at a particular locality after harvest of the crops, and find it necessary to make collections from a farmer's store. This is often important for vegetatively-propagated crops. Furthermore, it may be necessary to make collections from markets although the amount of information one obtains in markets can be quite small compared to that obtained directly from farmers. Some varieties may never be sent to market, and are only used for home consumption. Consequently, it is very important that as much as possible, even the remotest regions are explored as it is never certain what wealth of diversity might be found! Indeed, as genetic erosion continues, it is in these remote areas where the last remnants of diversity in a crop may still exist. Wild plants often grow in small disjunct populations, and in such circumstances the collection of large samples would most likely wipe out the species in certain localities. Obviously this should not be the intention of plant genetic resources collectors.

The distribution of collection sites must be governed by climatic and ecological considerations, and where these are highly varying, then collections should be made more frequently. When planning a collecting mission to an area, it is desirable to have knowledge of the distribution of variation in the crop or crops to be collected. When this is not available, and when it is possible to collect in more than one year, coarse-grid sampling is recommended. In subsequent years areas where the crops have shown more variation can be re-visited. Obviously such decisions can only be made on the basis of evaluation of germplasm. Traditionally such evaluations have involved growing material in the field, after which time it may be too late, genetic erosion may have taken place, and the return to areas of interest will lead to the discovery only of modern varieties. Damania *et al.* (1983) have shown that polyacrylamide gel electrophoresis of cereal storage proteins, a technique whereby proteins are separated according to their net electrical charge and molecular weight, can be used to determine the relative variation in cereal landraces. On the basis of such information, areas where diversity is greater can be identified, and collecting plans made or modified accordingly. Such procedures are laboratory-based, require little plant material and are quite rapid.

What if there is conflict between collecting large samples at a site and collecting from more sites in a day? Covering a larger area will capture more diversity associated with adaptation to geographic and microgeographic differences in environmental factors. So if there is a conflict resulting from the lack of time it is better to expend more effort on sampling more sites per day while collecting fewer plants per site.

Seed crops also differ in the ease with which they can be collected. The grasses and cereals are easy to collect because of the many seeds on each inflorescence. At maturity, the seeds are dry and can be harvested easily. But with plants which produce fleshy fruits such as tomatoes, chilli peppers, and melons, the extraction of seeds is both tedious and messy in the field. Such activities are best left for the evening back at the overnight hostelry or base camp. Nevertheless, seeds must be dried in order to obtain samples in a good physiological condition, and in addition it is not convenient to carry soft fruits for long distances during which time they can be damaged and decay.

In the countries of the European Economic Community, the introduction of legislation defining the crop varieties which may be grown has led to the elimination of many popular varieties from commercial agriculture. Often these varieties can only now be found on a local basis, preserved by keen gardeners, or locally available from some seed merchants. Crisp and Ford-Lloyd (1981) have advocated buying seed samples of vegetables from such seed merchants, and indeed on this basis many valuable collections have now been made in several countries.

Field documentation

As will be seen in Chapter 5, the continuing usefulness to the plant breeder of any germplasm accumulated during the exploration phase of genetic resources activities, will depend to a large part upon the information accompanying those

Expedition Organization: ....
Country: ....
Team/Collector(s): .... *Collector's Number:* ....
Date of Collection: .... *Photo Number(s):* ....
Species Name: ....
Vernacular/Cultivar Name: ....
Locality: ....
....
....
Latitude: ....°.......' *Longitude:* ....°.......' *Altitude:* ....
Material: Seeds Inflorescences Roots/tubers Live Plants Herbarium
Sample: Population Pure line Individual Random Non-Random
Status: Cultivated Weed Wild
Source: Field Farm Store Market Shop Garden Wild vegetati
Original Source of Sample: ....
....
Frequency: Abundant Frequent Occasional Rare
Habitat: ....
....
Descriptive Notes: ....
....
....
....
Uses: ....
....
Cultural Practices *Irrigated* *Dry*
Season: *Approximate sowing dates:* ....
Approximate harvesting dates: ....
Soil Observations: *Texture:* ....
Stoniness: ....
Depth: ....
Drainage: ....
Colour: ....
Soil pH: ....
Land Form: *Aspect:* ....
Slope: ....
Topography: Swamp Flood Plain Level Undulating
Hilly Hilly Dissected Steeply Dissected
Mountainous Other (specify)
Plant Community: ....
....
Other Crops grown near or in rotation: ....
....
Pests/Pathogens: ....
Name and Address of Farmer: ....
....
Taxonomic Identification: ....
by .... *Date:* ....
Name of Institution: .... *Accession No.* ....

Fig. 3.1 Example of general collecting form. (From Hawkes, 1983.)

collections. This information will not consist solely of evaluation characteristics, but also of details about where the collection was made, on what date, and what identification numbers and letters were given to it by the collector. In Chapter 5 it is suggested that the identifying details of any germplasm collection will be of immense importance if this collection is moved around the international community during its utilization, in order to avoid confusion and duplication of this germplasm and its use.

Other information which needs to be noted at the time of collection may serve to identify the material taxonomically, although if not done with some degree of accuracy should be left until evaluation is carried out at a later stage in the proceedings. Furthermore information which can **only** be recorded at the time of collection will relate to the source of the germplasm, i.e. wild habitat, farmer's field, farm store, market etc. Similarly, it may be desirable to make notes about the local habitat conditions such as soil type and rainfall for instance.

The amount of information which can be recorded during collecting activities is very much dependent upon the time available. There is a minimum amount of information which **must** be recorded regardless of time. But other details may be regarded as being of only secondary importance by some, and to be kept to a minimum so that the time spent actually collecting can be maximized. Others argue that extra time spent recording additional information will save time later on by avoiding some aspects of evaluation. Clearly a compromise can and must be made in this respect.

Various designs of collecting forms, sheets or books have been used in different collecting missions (Fig. 3.1), some of which may be fairly specific for certain crop species. It may even be possible to design a form which can be used at a later date for ease of direct input of data into a database (Fig. 3.2).

General considerations

All that we have said up to now has been concerned with sampling techniques applied on arrival in the field. The establishment of collection priorities on an international or national basis can only be made taking into account very many factors. The development of efficient exploration programmes for a particular species depends, according to Marshall and Brown (1975) on two levels of objective decisions: (1) where significant gene pools of the species occur and where are they most threatened by extinction; and (2) the areas covered by previous exploration missions and their relative effectiveness as measured by the material held in existing collections. The state of genetic erosion of a crop, as referred to in Chapter 1, and the status of world collections of crops must determine priorities for collecting. IBPGR which sponsors many of the current collecting activities has drawn up lists of crops which are under grave threat of genetic erosion on a world basis. It is likely that national priorities are at variance with global ones, but nevertheless careful consideration should be given to all factors before embarking upon an expedition.

Once decided upon, the broad stategy of an expedition must be adhered to strictly. For instance, if the primary objective of the expedition is to collect wheat, then this should not change part way in order to collect another crop.

Accession Data

Code	Descriptor	Categories	Entry
A1	Accession No.		________
A2	Species Name		________
A3	Subspecies		________
A4	Variety		________
A5	Original Breeder/ Supplier		________
A6	Donor Name/Inst.		________
A7	Donor Number		________
A8	Year of Donation		____
A9	Source	1 wild 2 farmland 3 farmstore 4 backyard 5 village mkt 6 commercial mkt 7 institute 8 other	__
A10	Status	1 wild inland 2 wild maritime 3 weedy 4 breeders line 5 primitive cv. 6 advanced cv. 7 other	__
A11	End Use	1 leaf vegetable 2 root vegetable 3 leaf and root 4 fodder 5 sugar 6 none 7 mangel	__

Collection Data

Code	Descriptor	Categories	Entry
B1	Collecting Inst.		________
B2	Collection Number		________
B3	Collection Date		__ __ __
B4	Country		________
B5	Province/State		________
B6	Nearest Town etc.		________
B7	Distance (Km)		____
B8	Direction		____
B9	Altitude		____
B10	Population Area		____
B11	No. of Plants in Popn	1 <10 2 10–100 3 >100	__
B12	Popn Isolation	1 from wild popns 2 from cult. popns 3 from wild & cult. 4 near wild & cult.	__
B13	Popn Uniformity	1 uniform, one type 2 var., one type 3 mixed types	__
B14	No. of Plants Sampled		____

Fig. 3.2 Example of form specifically designed for beet germplasm passport data.

Some expeditions are multi-crop expeditions, but even during single crop expeditions it is possible to find the occasional opportunity to collect some samples of other species. Likewise if it has been decided to collect in certain mountainous areas, these should be searched thoroughly, with no change of plan to collect the plains where differences in microhabitats will not be so marked. However, the day-to-day decisions, the tactics of collecting, must remain on a flexible basis. We have both experienced problems during collecting for potatoes in the Andes of Peru and beets in Turkey when plans have had to be modified due to circumstances beyond our control, such as transportation problems or illness. Roads may be closed due to landslides, and we have spent many frustrating hours waiting for roads to be opened. Political problems which develop in an area of a country, (and unfortunately such problems are becoming more common in many countries), may cause a change in tactics. We have had colleagues chased by gun-toting bandits in the Middle East. Increasingly it is necessary to leave the main roads, along which many collections have been made in the past, and travel into remote areas, either on foot or horseback. However these difficult journeys into more remote regions can be extremely rewarding in terms of plant genetic resources, and are worth the considerable effort to augment the samples obtained.

4
Conservation

The strategy of conservation depends on the nature of the material and on the objective and scope of the activity. The nature of the material is defined by the length of the life cycle, the mode of reproduction, the size of the individuals, and its ecological status, whether wild, weedy or domesticated (Frankel, 1970). The purpose to which the material is put will determine the degree of integrity which it is essential or desirable to maintain. Conservation strategy must take into consideration the time dimension; whether it be for short-, medium- or long-term storage, and where the storage will be located. The genetic make-up of the material and the type of sample collected will reflect the breeding system of the material, and how it will be regenerated.

There are two basic approaches to germplasm conservation, namely *in situ* and *ex situ* conservation, and these will be discussed separately.

In situ conservation

This is an approach which is applied mainly to wild species related to crop plants, to forest and pasture species. It is often recommended that these species should be preserved, maintaining the genetic integrity of their natural state, as communities in stable environments. The establishment of natural or genetic reserves recognizes the long-term objectives and the need for continued evolution within natural environments. Most nature conservation programmes are aimed at the level of ecosystems, or multispecies communities. Genetic conservation goes further in recognizing the need for a wide genetic base, and as with nature conservation, aspects of population biology such as adaptive radiation, maintenance of variation in populations, and the long-term stability of population numbers are important features.

There are areas in Israel, for instance where diversity in wild wheats, barleys, oats and *Aegilops* is great, where they grow among rocks and on poor soils and where the rough terrain creates a natural sanctuary. These communities are now protected against heavy grazing pressure from sheep and goats, and can continue to evolve. In Anatolia, wild orchards of pears, apples, plums and pistachio have been preserved for thousands of years, but are now under threat with the development and spread of agriculture. Such orchards will only survive if natural reserves are created.

The situation of the primitive cultivars, or landraces is very different from that of wild species. These categories evolved in association with humans, in the agricultural environment, or agro-ecoystem. Such systems are continually changing, and landraces are being rapidly replaced by modern cultivars. This process of 'genetic erosion' has been under way for decades, and for some crops such as wheat and barley, many of the ancient landraces have vanished. It has been suggested by some workers that small areas should be preserved in which landraces are cultivated according to traditional agricultural methods, so that the dynamic evolutionary processes can continue within the crop. However, the adoption of such a policy would condemn certain sections of communities to social and economic stagnation in a fast changing world, and as such is not often politically acceptable. The application of *in situ* conservation of landraces is not, therefore, a feasible option, and because of the rapid loss of genetic variation through cultivar replacement, *ex situ* conservation in one form or another is the most practical and safe approach for such material.

Ex situ conservation

In the broad sense this form of conservation can include the use of botanic gardens and arboreta on the one hand and genebanks on the other. Botanic gardens and arboreta represent the oldest forms of conservation, and in Europe date back to the 15th and 16th centuries. Little material has survived since then however, apart from a few trees, so that most are in fact of later origin.

The use of such vehicles for conservation has been considered more recently and extended to include the use of mass reservoirs as proposed by Simmonds (1962). He suggested the use of composite crosses of large numbers of diverse parental types as more appropriate for long-term conservation than what he called 'museum collections' of individual accessions kept separately, as would apply certainly in botanic gardens. The question as to the validity of their use is whether genetic variation is maintained more effectively in such reservoirs than in collections of individual accessions. Much evidence is available now from population genetic studies to indicate that loss of genetic variation on a large scale is in fact a reality. Marshall and Brown (1975) have estimated a genetic loss of between 50% and 70% in a barley composite cross kept in a mass reservoir.

Guldager (1975) has distinguished four kinds of *ex situ* conservation. These are appropriately illustrated when considering forest genetic resources (see Chapter 8). But for now let us say that two of these cover the models which have so far been described, and include 'evolutionary conservation' which implies the initiation of new natural evolution (natural selection in the environment of adoption), and 'selective conservation', which implies the application of selection pressure against clearly undesirable characteristics, without undue depletion of the available diversity. Another is that of 'static conservation' which is aimed at retaining as far as possible the population structure of the original population. Here, Guldager distinguishes between conserving genotype frequency and gene frequency, the former only being truly achievable in vegetatively-propagated material. The aim is to prevent the loss of genetic information. Static conservation mainly takes the form of the storage of seeds or vegetative material in genebanks. Conservation carried out in this way by reducing the life processes to a low level is both the safest and the cheapest method.

The role of genebanks

Both FAO and IBPGR have defined two types of conservation centres: long-term base collections and medium-term active collections. Base collections will be held in genebanks, solely for 'posterity', and will not be drawn upon except for viability testing and subsequent regeneration, or if seed is urgently required of any accession which cannot be acquired from any other source. In medium-term active or working collections, samples can easily be sent to scientists working in other centres or issued to breeders in general.

Although a number of genebanks have been established in the tropical developing countries in the last 20 years, large collections are also held in the developed countries of the temperate region. It is not surprising, therefore, that germplasm conservation of temperate cereals is much further ahead than that of most tropical cereals, pulses and root crops. The number of accessions of temperate wheats held in long-term storage in North America, Europe and the Soviet Union for example, is over 150 000. Collections of sub-tropical wheats are more modest and are held in medium- or short-term storage. CIMMYT maintains a working collection of 50 000 wheat and triticale accessions, while ICARDA holds some 17 000 wheat samples, and the N.I. Vavilov Institute in Leningrad has 70 000 wheat accessions including landraces. Breeders evaluating and utilizing sub-tropical and tropical wheats rely heavily on the small grains collection at Beltsville, Maryland and to a lesser extent on the National Seed Storage Laboratory in Colorado, USA.

IBPGR has been in the forefront of genetic resources work, particularly through the establishment of a 'global network' of genetic resources centres. This network includes national, regional and international institutions working to preserve the world's dwindling genetic resources. Among these institutions are International Agricultural Research Centres, one of which has been mentioned above (Table 4.1). These, and other designated institutions have agreed to be responsible for maintaining major base seed collections for the principal food crops. Further examples of these institutes are the Germplasm Institute in Bari, Italy, and the Nordic Genebank in Sweden which are coordinated on a regional basis, and the Greek Genebank in Thessaloniki and the National Vegetable Research Station genebank in the UK which are nationally organized.

Once a genebank has been established, and provided that it can satisfactorily fulfil the requirements of medium- and long-term storage of seeds or other organs, then its value will depend on the strategy that researchers follow with respect to collection, evaluation and provision of material for utilization. Within each of these three activities, documentation must play an important part, as the ultimate effective use of germplasm is totally dependent upon the availability and quality of information associated with it.

Table 4.1 International Agricultural Research Centres that have been designated by IBPGR as base centres for particular crops. (From CGIAR, 1982.)

Centre	*Crop*	*Nature of collection*
AVRDC	Mungbean (*Phaseolus aureus*)	Global
CIAT	Beans (*Phaseolus*)	New World
CIP	Potato	Global, wild and cultivated species
ICARDA	Barley	Global
	Chickpea	Global
	Faba bean (*Vicia faba*)	Global
ICRISAT	Sorghum	Global
	Pearl millet	Global
	Minor millets (*Eleusine, Setaria, Panicum*)	Global
	Pigeon pea	Global
	Groundnut	Global
	Chickpea	Global
IITA	Rice	African
	Cowpea (*Vigna unguiculata*)	Global
	Cassava (*Manihot esculenta* clones and seeds)	African
	Sweet potato	Global (clones and seeds)
IRRI	Tropical rice (*Oryza* spp., *indica*, *javanica*, wild species)	Global

Storage of seeds

In terms of static conservation, the most convenient way of maintaining plant germplasm would be by storing seeds. The main exceptions to this are plants which are normally vegetatively-propagated and do not produce viable seeds (e.g. banana), and crops where the seeds produced are very short-lived (e.g. cacao, see Chapter 3). The latter are often referred to as being 'recalcitrant' because they will not stand drying below some relatively high moisture content without very serious loss of viability (Roberts, 1973). Fortunately, the majority of crop plants do produce seeds, and these show 'orthodox' behaviour, being tolerant of a decrease in moisture content coupled with temperature, so allowing storage for relatively long periods. The storage potential of orthodox seeds is influenced by inherent as well as external factors. Genetic differences may exist at the level of genus or species, but even cultivars may vary in their storage characteristics. The Leguminosae is a family of plants with hard, impermeable seed coats. Consequently, they are renowned for their longevity. For instance, Harrington (1972) records *Cassia* seeds as surviving for 158 years, and *Trifolium* for 100 years. There are even *Lotus* and *Lupinus* seeds thought to have survived for over 500 years. But not all Leguminosae species are long-lived: the peanut, for instance, is notoriously short-lived. Short-lived species of other families include onion, parsnip and lettuce.

Barley, which is not a hard-seeded plant, has been reported to have survived for 123 years under special conditions, but while it does not present any major problems of storage, much variation in storage potential exists between cultivars stored under the same conditions. Significant cultivar differences in longevity have also been reported in peas, water melon, cucumber and maize.

Another factor which may affect storability of seeds is the persistence of fruit or floral structures. It seems that seeds of many cereals or grasses store better if kept with their 'hulls' or glumes intact rather than if threshed, and this may be due to an inhibitory effect upon fungal growth. Threshing may also have an adverse effect on viability because of possible mechanical damage during processing. Roberts (1972) has discussed this point and suggests that smaller seeds suffer less in this respect than do larger ones. Seeds which are round are also less likely to be damaged before storage, than ones with more irregular shapes.

When considering storage of seeds collected during expeditions, one factor which is difficult to account for is seed maturity. Scientists have regarded seed maturity as being that stage of maturation at which maximum dry weight has been attained (Roberts, 1972). Many crop species produce mature seeds over periods of days or even weeks, and thus may create difficulties of standardisation during collecting. In general, it is better to store fully mature seeds than immature, although in a few cases the degree of maturity does not seem to affect the storage potential.

Various pre-harvest and post-harvest extrinsic factors may have a bearing upon the life-span of seeds in store. Seeds subjected to moisture or temperature extremes during maturation, harvesting and processing may exhibit reduced viability thereafter. Clearly, the weather has an important role to play here. Strong correlations between weather conditions during ripening and harvesting, and loss of viability have been found in several crops such as barley, wheat and oats. While it is important to appreciate the effects of such environmental factors, germplasm preservation has to go on regardless of the weather!

Factors controlling longevity during storage

When storing seeds in a genebank it is necessary to control certain parameters in order to maximize the longevity of the seeds. Much is known about the requirements for seed stores and the two most important factors are seed moisture content and temperature. It is difficult to discuss these parameters separately. Indeed Harrington (1963) has indicated that, roughly speaking, longevity is doubled for each 5°C fall in temperature or for each 2% drop in moisture content. This is Harrington's 'rule of thumb'. From this, it might be expected that when seeds become fully hydrated again, such as they would in the soil, their potential period of survival would become extremely short. In general, this does not seem to be the case, for when moisture content reaches a level of about 18%, Harrington's rule breaks down, and increasing moisture content no longer affects longevity as dramatically. Even more interesting is that when the seed is close to full hydration, an actual increase in longevity occurs providing the seed remains dormant (Villiers, 1974). Roberts and King (1982) report this to be quite accentuated in lettuce for instance. It also partly emphasizes why annual weed seeds, generally regarded as being orthodox in their behaviour, are often capable of surviving for relatively long periods in the soil.

It is possible to dry seeds of many species to a 6% moisture content without damage, but some kinds are injured by drying to lower moisture levels. IBPGR now makes recommendations along these lines for the preferred storage conditions for the long-term conservation of orthodox seeds (IBPGR, 1976). It is suggested that seeds should be dried to 5 ± 1% moisture content, placed in sealed containers and stored at ⁻18°C or less. By storing seeds under such controlled conditions, it is thus possible to calculate the longevity of seed lots in terms of the remaining viability after a certain storage period (Roberts, 1973), using a complex procedure of three viability equations incorporating four viability constants.

Although storage under optimal conditions will result in the maintenance of high viability over a long period, a certain decrease in viability is inevitable even under such conditions. How does this relate to the lifespan of individual seeds? Within a single, genetically homogeneous population stored within a stable environment, the lifespans of individual seeds differ considerably. The frequency of deaths per unit time conforms to a normal distribution. The standard deviation is then increased in better storage conditions (lower temperatures and/or moisture contents) since the lifespan of every seed is increased by the same proportion. Following on, it is possible to calculate survival curves for different seed lots. The shapes of these will be identical for different seed lots of any given species, stored under the same conditions but these seed lots may differ considerably in the time taken to fall to a given level of viability, since the survival curves may be displaced in time as a result of differences in genotype and pre-storage conditions.

Because the survival curves are not linear, and because seed lots differ in the extent to which they have deteriorated before storage, the change in percentage viability over any given storage period differs between seed lots. Fortunately, however, it is now possible to estimate this pre-storage deterioration and subsequent rate of loss of viability. This allows for the calculation of how long it will take any seed sample to fall below a certain level of viability and therefore when it should be considered for 'regeneration' (grown to supply fresh seed for further storage), as shown in Table 4.2.

Table 4.2 Estimates of probable regeneration intervals for seeds stored at −20°C and 5% moisture. (Modified from Roberts, 1973.)

Species	*Cultivar*	*Probable regeneration interval (years)*
Barley (*Hordeum vulgare*)	Proctor, Golden Promise, Julia (mean value)	70
Rice (*Oryza sativa*)	Norin	300
Wheat (*Triticum aestivum*)	Atle	78
Broad bean (*Vicia faba*)	Claudia Superaquadulce	270
Pea (*Pisum sativum*)	Meteor	1090
Onion (*Allium cepa*)	White Portugal	28
Lettuce (*Lactuca sativa*)	Grand Rapids	11

Recent research has indicated that sequential germination tests are as reliable as fixed sample size tests, while at the same time requiring considerably fewer seeds to determine when seed regeneration should take place (Ellis and Wetzel, 1983). Such a monitoring test would comprise a sequence of an indeterminate number of germination tests on small groups of seeds sampled from an accession, and requiring the sacrifice of far fewer seeds than other procedures.

Genetic change during storage

Apart from the possibility of alteration of the genotype composition of a seed lot during storage because of differential loss of viability between genotypes, there also remains the possibility of chromosomal genetic damage. Although such nuclear damage may not be the cause of loss of viability, breakage of chromosomes unable within the seed to undergo normal DNA repair processes, and consequent build-up of recessive mutations will have important consequences with respect to the problems of genetic conservation.

One answer to this is, of course, to regenerate seed. Whether or not this is a particularly satisfactory answer is questionable. As will be seen later, the regeneration process itself poses problems in terms of maintenance of genes within collections, and these problems may certainly outway those imposed by the accumulation of mutations. Another answer is freezing at very low temperatures. This has been carried out for various seeds without any apparent adverse effects upon germination percentage or seedling development (Leron Robbins and Whitwood, 1973; Sakai and Noshiro, 1975; Mumford and Grout, 1978, 1979). The storage time for such experiments has never been very long, and never more than 60 days, but as the chemical reactions that constitute normal cell damage are totally inhibited at the low temperatures used (⁻192°C or ⁻196°C), the potential must exist for storage over many years (Mazur, 1966). The only damaging effects likely to occur at this temperature are those of direct ionization on macromolecules such as DNA and it has been predicted that decades would have to elapse before a statistically significant level of mutation resulted from background radiation (Mazur, 1976).

Seed storage facilities

Generally, on a large scale, seed is stored in some form of controlled environment room, where as a minimum requirement the temperature can be maintained at ⁻18°C or less (Fig. 4.1). The humidity of the room may also be regulated to a level of roughly 5% RH. This, however is technically difficult and therefore expensive, and consequently many genebanks store seeds in hermetically sealed containers. These may take the form of glass vials, metal cans or laminated aluminium foil packets. The seed samples, already dried to a low moisture content, are placed within these containers which are then sealed, the operation being carried out in an atmosphere of 5% RH.

Technical details of the design and construction of seed storage facilities are important but complex, and the level of efficiency must often be counterbalanced with cost (Cromarty *et al.*, 1982). For storage of seeds on a much smaller scale, domestic 'deep-freeze' chests are perfectly adequate, particularly if pre-dried seeds are kept in one or other form of sealed container. Although the storage of all germplasm under the conditions outlined is desirable, the storage of more short-term or active collections is often less stringent for purely practical and economic reasons.

Clearly the storage facilities described are dependent upon a fairly constant energy supply (electricity) for reliable and effective medium- and long-term storage. Occasional brief disruptions of power are likely to happen even in the best of circumstances, and these can be tolerated with little worry about loss of germplasm viability in the long term. After all, a controlled temperature room is likely to take at least a few days to reach ambient temperature, which in itself will not be harmful.

Only if prolonged disruption (months) to the power supply is envisaged will measurable permanent damage be done. Such an occurrence may arise from civil disturbance or extreme political activity, in which case it may be worth considering the proposals underway at the Nordic Genebank (Yngaard, 1983) to preserve duplicates of seed samples in permafrost in a remote cave in Norway.

Fig. 4.1 A typical long-term seed storage facility, containing hermetically-sealed aluminium laminated packets, maintained at −20°C. (© National Vegetable Research Station.)

Problems with recalcitrant seeds

Many species of fruit and some of the large-seeded tree species have been described as recalcitrant since, as the term implies, they do not obey the 'rules' which can be applied to orthodox seeds. Even under moist storage conditions they are relatively short-lived and last no more than a few weeks or a few

Table 4.3 Species producing recalcitrant seeds. (From Withers and Williams, 1982.) This is a summary of a list which appears in King and Roberts (1980) but excluding *Citrus* spp. of which at least some are now considered to be orthodox and *Coffea* spp. about which there is some doubt. The original list gives references, details of storage conditions used, the maximum storage periods achieved with final percentage viability where known, and in some cases notes on conditions which are particularly damaging.

Species		**Storage period** (d = days;w = weeks; m = months;y = years) **and % final viability** (in brackets)
ACERACEAE	*Acer saccharinum* L. (Silver Maple)	16 m (75%)
ANACARDIACEAE	*Mangifera indica* L. (Mango)	13 w (80%)
ARAUCARIACEAE	*Araucaria excelsa* R. Br. (Norfolk Island pine)	5 m
BOMBACEAE	*Durio zibethinus* Murr. (Durian)	32 d (90%)
CORYLACEAE	*Corylus avellana* L. (European filbert)	6 m +
DIPTEROCARPACEAE	*Dryobalanops aromatica* Gaertn. (Borneo camphor)	20 d (30%)
	Hopea helferi	2 m
	H. odorata	3 w (89%)
	Shorea ovalis	3 m
	S. talura	6 m
EBENACEAE	*Diospyros kaki* L. (Japanese persimmon)	18 m (96.5%)
ERYTHROXYLACEAE	*Erythroxylon coca* Lam. (Coca)	30 d (2%)
EUPHORBIACEAE	*Aleurites fordii* Hemsl. (Tung)	54 m (35%)
	A. montana (Lour.) Wils. (Tung)	6.5 m (81%)
	Hevea brasiliensis (Willd.) Muell.-Arg. (Rubber)	3.5 m (100%)
FAGACEAE	*Castanea crenata* Sieb. *et* Zucc. (Japanese chestnut)	ca. 6 m
	C. mollissima Blume (Chinese chestnut)	ca. 6 m
	C. sativa Mill. (Spanish chestnut)	ca. 6 m
	Castanea spp.	< 3.5 y
	Quercus acutissima	2 m (61%)
	Q. alba L. (White oak)	ca. 6 m
	Q. borealis Michx. (Red oak)	41 m (36%)
	Q. coccinea Wagenh. (Scarlet oak)	ca. 6 m (90%)
	Q. pagodaefolia (Ell.) Ashe (Cherrybark oak)	6 m +

	Q. mongolica Fischer var. *grosseserrata*	2 m (60%)
	Q. nigra L. (Water oak)	6 m (73%)
	Q. petraea Lieblein (Sessile oak)	ca. 5 m (22%)
	Q. robur L. (Pedunculate oak)	42 m (70%)
	Q. suber L. (Cork oak)	8 m (86%)
	Q. velutina Lam. (Black oak)	5 m
GRAMINEAE	*Glyceria striata* Hitche (Manna grass)	7 m +
	Zizania aquatica L. (Indian rice)	14 m (86%)
GUTTIFERAE	*Garcinia mangostana* L. (Mangosteen)	6 w (100%)
HIPPOCASTANACEAE	*Aesculus hippocastanum* L. (Horse chestnut)	15 m (25%)
JUGLANDACEAE	*Carya pecan* (Marsch.) Engl. *et* Graebn (Pecan)	24 m +
	Juglans spp. (Walnuts)	6 m (18% – 92%)
LAURACEAE	*Cinamommum zeylanicum* Nees. (Cinnamon)	1 w (80%)
	Persea americana Miller (Avocado)	12 m (75%)
MALVACEAE	*Montezuma speciosissima* Sesse *et* Moc. (Maga)	1 m (50%)
MELIACEAE	*Swietenia* sp. (Large leaf mahogany)	4.5 m (72%)
MORACEAE	*Artocarpus heterophyllus* Lam. (Jackfruit)	1 m (80%)
MYRTACEAE	*Eugenia dombeyana*	—
PALMAE	*Cocos nucifera* L. (Coconut)	16 m
	Elaeis guineensis Jacq. (Oilpalm)	15 m
PROTEACEAE	*Macadamia ternifolia* F. V. Muell. (Macadamia nut)	11 m (8%)
ROSACEAE	*Eriobotrya japonica* Lindl. (Loquat)	6 m (92%)
SAPINDACEAE	*Nephelium lappaceum* L. (Rambutan)	1 m (100%)
STERCULIACEAE	*Cola nitida* Vent. (Gbanja kola)	5 m (80%)
	Theobroma cacao L. (Cocoa)	8 m (24%)
THEACEAE	*Thea sinensis* L. (Tea)	10 m (50%)
TRAPACEAE	*Trapa natans* L. (Water chestnut)	7 m (92%)
ZINGIBERACEAE	*Elettaria cardamomum* (L). Maton. (Cardamon)	—

months, depending on the species. It is quite clear that recalcitrant seeds have a physiologically distinct behaviour from orthodox seeds, and their delimitation rests mainly on their inability to withstand desiccation and their short-lived characteristics even when fully imbibed.

From the genetic resources point of view, recalcitrant crops of interest include such economically important species as cocoa, rubber, tea, most tropical fruits and many timber species (Table 4.3). The identification of recalcitrant types is however, not always easy: for instance, citrus species were once considered recalcitrant, but are now known to be orthodox (IBPGR, 1976; Mumford and Grout, 1979; King and Roberts, 1980).

This has recently been found to be due to the delaying effect that a dried testa has on the germination of fresh seeds. If the germination test is prolonged sufficiently, or if the testa is removed, then *Citrus* species are truly orthodox. The embryo of *C. limon* for instance, is capable of withstanding extreme desiccation without adverse effects on subsequent growth.

Problems of maintaining viability of recalcitrant seeds are not just related to storage for conservation. Often, such seeds are so short-lived that major problems arise even during actual collections. The little progress which has been made in the storage of truly recalcitrant seeds has been towards keeping the seeds viable long enough for them to survive the period of collection. In most cases the best recorded methods involve storage in a fully imbibed state, in a moist medium in aerobic conditions (a loosely-sealed polythene bag), where the seed has received a preliminary treatment with hot water or a fungicide to inhibit microbial growth. Under such conditions, however, germination has often to be suppressed, either by lowering the temperature of storage very slightly, or by osmotic inhibition.

In order for storage of recalcitrant seeds to be of any value in conservation, other than during collection, then the storage period attainable needs to be longer than the minimum life-cycle of the plant from sowing to first harvest. This may of course be several years or even decades – periods of time which are not achieved by present methods of recalcitrant seed storage. As a result, there has been speculation as to the use of more novel techniques such as cryogenic preservation of embryos, by adapting the methods used successfully for the storage of animal tissues. Observation of the natural cycle within tropical forests may also provide a lead to further possible techniques. In the wild, tropical tree seedlings remain in their juvenile state for prolonged periods under the reduced light regimes prevailing in tropical forests (Hawkes, 1982). These seedlings retain their viability and potential for further development until disruption of the leaf canopy takes place bringing about an increase in available light. Could the maintenance of seedlings under such very low light conditions be a possible means of conserving recalcitrant species?

Seed regeneration

Apart from the need to regenerate seed to maintain viability in collections, the other purpose of regeneration is to provide an appropriate amount of seed for long-term conservation, as well as to restore the seed in active collections used for evaluation and utilization. The aim should be, as far as possible, to maintain as many of the original genes as possible within the sample. This point is somewhat controversial in the sense that some people might argue for the maintenance of the genetic structure of the original population in terms of the number and proportion of genotypes. The conservation of genotypes is of course important from a practical point of view in such crop examples where controlled genetic recombination during the plant breeding process may be difficult due to the inherent genetic system. But generally speaking, the conservation of every individual genotype would be an impossible task, 'gene' conservation being a rather more practical proposition.

Certain requirements need to be met. Firstly, a knowledge of the breeding system will allow for control to be effected to reduce outcrossing with other entries to a low level. Further, in cross-pollinating species in particular, fertilization may be controlled within accessions, perhaps to the extent of pairwise crossing, in order to double the effective population size. Secondly, there is a need to reduce the effects of natural selection acting in any environment other than the original one. Survival should be maximized so that all or most of the components of a population survive and reproduce.

Allard (1970) has stressed the major effect of genetic drift in small populations of either inbreeding or outbreeding species, leading towards gene fixation or genetic erosion. The loss of rare genes through genetic erosion can be reduced by considering carefully the number of plants and strategy required for regeneration. Frankel (1981) states that the effective population size for regeneration should correspond to that recommended for sampling in the field and under ideal circumstances should be a minimum of 100 plants. While it is likely to be quite impossible to avoid the loss of all rare genes or alleles due to practical constraints, knowledge of biometrical and population genetics should be applied to minimize this loss. Also, at every stage of the genetic consevation process, particularly during collection, evaluation and the regeneration of seeds, genes recognised as being important should be spotted and isolated for immediate fixation.

The problem of vegetatively-propagated crops

The conservation of vegetatively-propagated crops such as potato, cassava, yams, sweet potato, sugar-cane and many temperate fruit trees amongst others presents special problems. Although some of these crops are sexually fertile, it is often not convenient to propagate them commercially from seed because of high levels of heterozygosity, and breeders and horticulturalists commonly require uniform clones. Nevertheless the sexual potential of some of these crops does offer the possibility of an alternative conservation strategy as for any sexually-propagated crop.

Many vegetatively-propagated crops are, however, sexually sterile, or at the very least have reduced fertility which precludes the possibility of seed storage. There are many factors which reduce fertility, including genetic and chromosomal changes. Many of these asexual crops are high or odd-numbered polyploids combining genotypes from a variety of sources. They only survive because of their asexual nature.

The vegetative organs which are stored, namely tubers, rhizomes, corms and cuttings etc., are relatively short-lived and often deteriorate rapidly after harvest, unless ideal storage conditions are provided. The cultivation of collections vegetatively year after year is costly and brings with it dangers of disease infection and many other hazards. New methods of long-term storage of vegetative organs are urgently required, and some developments are given later in the section on *in vitro* storage. Storage of vegetative organs of tropical versus temperate crops is quite different. The tubers of yams and the tuberous roots of sweet potato store best at low temperature, but at temperatures lower than 14°C, there is tissue decomposition. High temperatures, while beneficial in the early stages of storage for hastening the physiological processes of curing in wound healing, promote other kinds of deterioration, including increased sprouting. High humidity is necessary during storage. Cassava is propagated by fresh stem cuttings and additional research is needed to prolong storage of cassava 'sticks' for up to a year. The variables requiring testing include levels of humidity and moisture, as do the methods of avoiding fungi, and the chemical inhibition of sprouting.

Potato tubers can be stored adequately for 5–7 months at 4–5°C and 90% relative humidity. There is considerable evidence that storage in diffused light compensates for higher temperatures (CIP, 1978). Sprouts develop relatively slowly and sturdily under such conditions. Greening of seed tubers is not a problem because they are not destined for human consumption, and the greening process may well confer some resistance to diseases and insects.

In all these cases it is important to realise that all the organs for storage are metabolically active. Their life spans follow closely controlled cycles associated with the annual cycle. The storage methods indicated here only represent short- or medium-term strategies. Annual regeneration is not without risk, in that germplasm is exposed to both the rigours of weather and disease, and some genotypes may well be lost. For long-term storage other methods must be adopted. It is often convenient to maintain large collections of vegetatively-propagated crops in their centres of diversity. Among the advantages are physiological and phytopathological aspects. Many crops are tropical in origin, and will only produce vegetative organs under strict photoperiodic regimes. Such organs are bulky, and difficult to transport. They are easily damaged in transport, and this can lead to losses. But perhaps of more importance are the quarantine and disease control aspects. It is extremely difficult to keep vegetatively-propagated plants free from viruses, and these represent a major phytopathological threat. Infection with virus diseases leads to degeneration of clonal stocks. Only a few virus diseases are transmitted sexually through pollen and true seed, and most can be eliminated by propagation of a crop through true seed. Unfortunately this is not always possible because of the lack of fertility, nor always desirable because of the need to preserve particular genotypes. Fungal and bacterial diseases are more easily controlled by appropriate measures.

The maintenance of large living collections requires large inputs of labour and land. The conservation of fruit trees, usually in the form of orchards, is beset by many problems. These include space considerations associated with the need to conserve an adequate sample of genetic variation. Sykes (1975) has indicated that sampling only 0.1% of the estimated three million almond trees in Turkey would provide 3000 single-tree accessions, occupying some 15 ha, but grafting different accessions onto the same root stock, or using dwarfing root stocks would reduce space requirements. In addition, there are numerous other fruit tree species in need of conservation. Nevertheless, the conservation strategy for such species is daunting.

With these living collections and indeed with crops which are regenerated annually, documentation and labelling are extremely important. Errors in these may lead to mixtures of clones, and detract from the value of detailed records of clone morphology.

While the conservation of crops such as potatoes has progressed well, with a clonal genebank at CIP in Peru, as illustrated in Fig. 4.2, and seed collections maintained also in the United States, the United Kingdom and Germany, the situation in other crops is critical. If genebanks are not established quickly, there is a definite danger of losing forever much of the variation in these crops.

Fig. 4.2 Part of the clonal potato genebank maintained by the International Potato Centre at Huancayo in central Peru, at 3300m. The collection has been reduced from 13 000 to 5000 accessions through the identification of duplicate clones. The collection is planted annually. (Courtesy of the International Potato Centre.)

In vitro storage of germplasm

Much interest has been focussed in recent years on the application of tissue culture or *in vitro* techniques to plant genetic conservation because of the special problems presented by vegetatively-propagated and recalcitrant species. The basic aim of such methods is to introduce explants from these plants into sterile culture and maintain them in a pathogen-free environment, making them available for future use. In addition, the introduction of such cultures to ultra-low temperatures then opens the possibility of storing germplasm indefinitely, in 'suspended animation' (Henshaw, 1984).

The detailed techniques of tissue culture have been admirably covered elsewhere (Dodds and Roberts, 1982; George and Sherrington, 1984), so that it is our intention only to discuss some of the special problems which relate to the storage of germplasm using *in vitro* systems. Many methods have been developed for the propagation of different species, including callus, meristem, and anther culture, as well as the maintenance of isolated protoplasts, but there are considerable deficiencies, nevertheless, in our understanding of these systems which have serious implications for genetic conservation. Until it is possible to regenerate this material with a high degree of success, then there is always the danger that germplasm will be lost during the culture phase.

There are two major aspects when considering tissue culture for plant genetic conservation, namely: (*i*) the techniques to be used which can be either normal and minimal growth in culture, or freeze preservation, often referred to as 'cryopreservation'; and (*ii*) the problem related to genetic stability. It is convenient to deal with each of these individually, although they are interrelated to a considerable extent.

Normal and minimal growth in storage

Tissue cultures can be maintained under normal growth conditions virtually indefinitely provided that nutrients are supplied and accidents are avoided. *In vitro* systems with a high multiplication rate are not suitable, however, for germplasm conservation. Such systems require frequent attention and maintenance. But perhaps the most important aspect is that in unorganized systems such as callus cultures, mutation rates are likely to be high because of the high rate of cell division. Minimal growth strategies in which shoot-tip cultures and plantlets from meristems grow at very slow rates, will undoubtedly have further application in genetic conservation. Clearly these methods of storage have considerable advantages in that the stored material is readily available for use, can be easily seen to be alive, and cultures may be replenished when necessary. Methods used to induce minimal growth may involve the following.

(*a*) An alteration of the physical conditions of culture, most often by a reduction of the temperature. Generally, tropical crops cannot tolerate a temperature reduction as well as temperate crops. For example, crops such as potatoes, apples, strawberries and grasses can be stored at 0–6°C, whilst crops such as cassava and sweet potato are stored within the range 15–20°C.

(*b*) Alterations to the basic medium, by the omission or reduction of some factor essential for normal growth. This has been done routinely with apples and plums, for example.

(*c*) Use of a growth retardant, such as abscisic acid, or compounds with osmotic effect such as the sugar alcohols mannitol and sorbitol.

Particular success has been achieved with potatoes (Westcott, 1981a,b), and rosaceous fruit trees (Wilkins and Dodds, 1983) using one or a combination of the above methods, and a germplasm collection of the first crop is partially maintained *in vitro* at CIP. These methods are applicable for short- or medium-term storage, but for long-term storage the elimination of routine sub-culturing is desirable.

Cryopreservation

The freeze preservation of cultured animal cells, spermatozoa and ovarian and embryonic tissues, as well as whole animal embryos has had a history of several decades compared with the research on plant tissues. The potential for the long-term storage of plant germplasm using freeze preservation is vast, although the application of cryopreservation techniques is still in its infancy, and reliable, routinely-usable procedures for organised vegetative tissues are not yet available for any major crop (Keefe and Henshaw, 1984).

During the freeze preservation of plant tissue there is a complete cessation of cell division; it is important that damage by ice crystal formation within the cells is prevented altogether or at the very least, minimized, as the temperature of the culture is reduced to that of liquid nitrogen (⁻196°C). Two methods have been developed to achieve this, namely the ultra-rapid freezing and the slow, stepwise freezing of cultures. In the former, ice crystals form within the cells, but are extremely small and therefore cause no disruption. Cells must be rapidly thawed to prevent recrystallization; this method has been successfully applied to potato shoot-tips. The slow freezing method depends for its success on extracellular freezing for protection, and this is the method which has been most widely and successfully applied. In both systems, however, the tissue is often treated with a cryoprotectant such as DMSO (dimethyl sulphoxide) or glycerol which reduces damage in a number of ways. Cryoprotectants have the ability to reduce the size and growth rate of ice crystals, and lower the freezing point of intracellular contents and enable cells to be subjected to very low temperatures without disruption of the cell membrane or contents. Recent research by Keefe and Henshaw (1984) has shown that artificial nucleation of the cryoprotectant medium (DMSO) at a temperature close to its freezing point has a beneficial effect, and explants of *Solanum tuberosum* suffered less freezing injury when this occurred.

Most cryodamage is related to membrane structure and function; in contrast, nucleic acids are very resistant to freezing and thawing. Once frozen the samples may be transferred to a liquid nitrogen refrigerator for long-term storage. A variety of thawing methods may be employed to defrost samples, which are then placed in culture. As Wilkins *et al.* (1982) point out there is no reason to believe that long-term conservation of plant tissue cultures should not be achieved since animal cells have been stored in a similar manner for over 20 years, and have remained viable. Nevertheless, for plant tissues, an understanding of the operation of mechanisms of freezing injury is essential if cryopreservation is to fulfil its potential.

Although it is normal procedure to freeze cultures, some research on various tree species, including *Salix*, *Populus* and *Betula* and a wide range of fruit-trees, including *Malus* (apple) has indicated that it is possible to freeze twigs in liquid nitrogen, after passing them through a pre-freezing treatment (Sakai, 1965; Sakai and Nishiyama, 1978). Twigs can then be thawed slowly and placed in water or sand to await bud development. Most success has been achieved with hardy winter twigs.

The storage of imbibed seeds and zygotic embryos dissected from imbibed seeds is worthy of consideration since it may have potential application to recalcitrant seed. Such tissues have a high degree of actual and potential organisation and contain both root and shoot meristem regions. Withers (1982) has indicated that considerable difficulties have been encountered in the cryopreservation of embryos from plants such as maize and carrot which have been studied. Nevertheless, there is much potential in this approach.

Genetic stability

The ultimate aim of any of the tissue culture systems in plant genetic conservation is the long-term preservation of genotypes. Systems which are prone to genetic change are, therefore, less desirable. The maintenance of a consistent 'germ-line' through several generations requires an ordered process of chromosome replication in the cells of the apical and other meristem regions. Apical meristems are more favoured for storage and the propagation of species through tissue culture (Fig. 4.3). Callus cultures are prone to genetic instability because of their relatively unorganized differentiation. Genetic stability is likely to be maintained however where continued organized growth of the meristem in culture follows culture initiation.

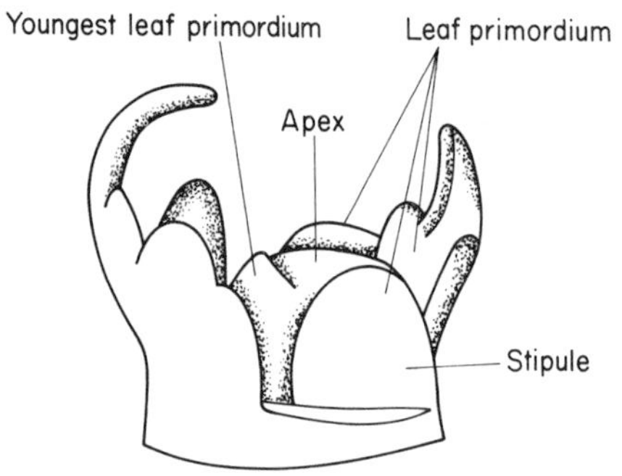

Fig. 4.3 Diagrammatic representation of an isolated meristem for culture. This has an apical dome, together with two or three leaf primordia. (From Wilkins *et al.*, 1982.)

Parent plants in which meiotic and other cytological abnormalities are common will often be more susceptible to genetic changes in culture. Increases in chromosome number, i.e. polyploidization, may result from spindle failure and chromosome lagging during mitosis, and these and other changes leading to odd chromosome numbers or even reductions in chromosome number have been reviewed by a number of authors (Bayliss, 1980; D'Amato, 1978). Certain tissue culture systems lose their potential to differentiate after growth under *in vitro* conditions for a long period, and some changes such as aneuploidy do increase in frequency with the age of the culture. The genetic integrity of a culture may be affected by different culture media. Some genotypes may be favoured at the expense of others.

Any tissue culture system must also guarantee the retrieval of material at the end of the storage period. The ability of cultures to differentiate, that is the morphogenetic potential, must be retained. Again, organized cultures such as those derived from meristems offer the most advantageous and simplest option.

When deciding upon a conservation strategy employing tissue culture systems, the relative advantages of each must be evaluated (Fig. 4.4). A system which provides optimum storage conditions may not be entirely appropriate for plant regeneration after storage, and vice versa. The genetic stability and potential for both quantitative losses and qualitative changes are all factors which will determine which tissue to utilize and the most appropriate storage conditions. Undoubtedly freeze preservation offers the best hope for long-term storage. Under such conditions there is only one source of genetic variation, namely mutation induced by background radiation. At ultra-low temperatures, healing processes will not occur and mutations may accumulate (Withers, 1980).

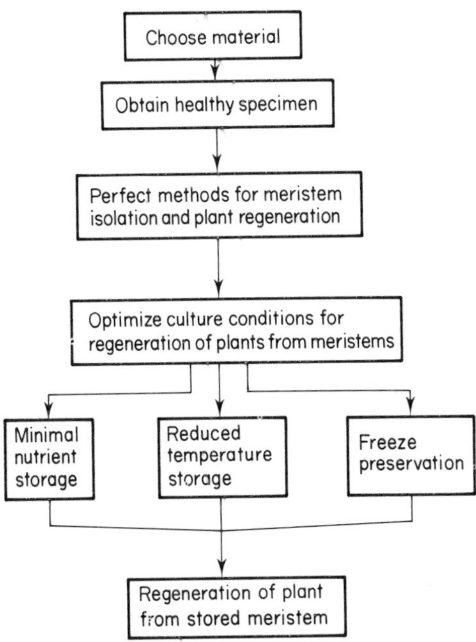

Fig. 4.4 Flow diagram of the strategy for conservation of any plant. (From Wilkins *et al.*, 1982.)

Further considerations

The application of *in vitro* methods of plant genetic conservation has several additional advantages. During the culturing of vegetatively-propagated crops, many serious diseases, particularly viruses, can be eliminated and disease-free stocks maintained (Henshaw, 1984). This has important implications for germplasm exchange and international plant quarantine. Where the aim is to eliminate systemic pathogens, such as viruses, the choice of apical meristems, together with a couple of leaf primordia as the explant is advisable. Such explants are frequently free of viruses even when these have been detected in the original plants. Furthermore, the chances of obtaining pathogen-free cultures are increased when the plants have been subjected to the thermotherapeutic treatments, which involve growing plants for several weeks at elevated temperatures in the range of 36–40°C. Such methods cannot always guarantee success, and research is underway to complement thermotherapy with chemotherapeutic treatments. Nevertheless, considerable success has been achieved with crops such as potatoes. The International Potato Centre exports only pathogen-tested germplasm as *in vitro* cultures (Roca *et al.*, 1979), and the same procedure is now being applied to cassava from CIAT (CIAT, 1981). Tissue culture also has a major advantage in that a large number of genotypes can be stored in a relatively small area and generally at a fraction of the cost of growing material annually in the field, or maintaining large living collections of fruit trees.

5

The Management of Genetic Resources Data

The conservation of rapidly disappearing genetic stocks for possible future use and their subsequent utilization in plant breeding are two highly important spheres of activity in any genetic resources strategy. But the success of these is dependent upon the availability of descriptive information which will enable plant breeders to make decisions regarding the material to be introduced into breeding programmes. Adequate information systems are also essential for the smooth operation of the global network of genebanks, while germplasm collections are more useful if reliable descriptions of their contents are easily accessible to breeders.

Magnitude of the problem

One major factor affecting genetic resources information handling and exchange is the sheer volume of existing germplasm collections and their associated descriptions. To obtain some idea of the magnitude of the problem of handling data on genetic resources, it is worthwhile considering some of the major stores of germplasm in the world today. For instance, IRRI in the Philippines holds over 63 000 samples of rice. The morpho-agronomic data alone run to over 38 traits recorded, giving something in excess of 2 million pieces of information to manage. Likewise, CIP stores data on 16 000 accessions of wild and cultivated forms of potato. These represent extensive evaluations for factors related to yield, resistance to diseases, viruses, nematodes and insect pests. They also include information on pedigrees and details of crosses, as well as of course collection information and taxonomic details.

The need for information retrieval

If the proposition is accepted that some germplasm collections have associated with them as much information as that contained within a good English dictionary, the only way of accessing information kept in such a form is alphabetically. If the spelling of a word is known, this will allow the meaning to be found. Retrieval of information in any other way is, to all intents and purposes impossible. For instance, if it was required to find all words related through similar meaning, reference would have to made to a different 'database' i.e. a thesaurus, card index or printed catalogue. These arguments apply just as much to genetic resources data. A card index or catalogue may contain information about accessions ordered strictly according to accession numbers or species for instance. They cannot be ordered both ways at the same time, except in a limited fashion by using sub-orders i.e. alphabetically by species first, and then by accession number within each species. Such a system would work reasonably well if a retrieval request was put in the form of 'please find me the accessions with accession numbers between 1702 and 1754, and which are of the species *vulgaris* or *maritima*'. The above system could not cope adequately with a similar request but including only those accessions collected in Turkey before the year 1975, for instance. When querying a large database in such a complex manner computerized information retrieval is essential, and could be effected by a single logically defined command such as 'FIND ME ACCESSIONS WHERE SPECIES EQUALS VULGARIS OR SPECIES EQUALS MARITIMA AND WHERE YEAR OF COLLECTION IS LESS THAN 1975 AND WHERE COUNTRY EQUALS TURKEY AND WHERE ACCESSION NO. IS BETWEEN 1702 AND 1754'.

Retrievals or manipulations of the information should not be thought of as being restricted to passport or evaluation data (see below for explanation), but should be extended particularly to genebank maintenance data and operations. For even small collections of genetic resources data computers provide the only effective answer to data management problems.

The form of descriptive information

Genetic resources work is about living plants. The information that we will need to handle will serve to describe these plants in terms of form, function and origin. It is these descriptions which are stored and processed in the computer. 'Descriptor' is now widely accepted as the computer term for the 'character' of a plant, as well as for other units of information such as the country of origin of the plant, or the date of its collection. The 'descriptor state' is then the quantity or quality of the plant character, or any country name or abbreviation, or the actual month or year collected.

It may now be apparent that from the point of view of use and content, several categories and types of descriptor can be distinguished. IBPGR has supported the use of four major categories of descriptor, three of which apply particularly for use in genebanks.

(1) *Passport information* – to consist of data recorded at the time of collection of germplasm, together with identifying names and numbers.

(2) *Characterization data* – descriptors for characters that are highly heritable, that can be seen easily by the eye and are expressed in all environments.

(3) *Preliminary evaluation data* – descriptors for a limited number of additional traits 'thought desirable by a consensus of users' of a particular crop. These should be capable of visual assessment, but may not be expressed in all environments (such as reaction to disease, drought or other stress).

Data falling into these categories are essentially 'historical'; this relates to frequency of data capture, which happens once only for any particular plant or group of plants.

(4) *Full evaluation data* – information on traits related in the main to breeding programmes.

The frequency of data capture of this fourth category is variable and may be unpredictable. It will often be recurrent; for instance when field trials are repeated over a number of years. It will be unpredictable because a plant breeder may decide for good reason to evaluate for a limited number of agronomic characters only, at any one time.

For the passport data, the descriptors falling into this category may be considered the easiest and most desirable to identify and standardize from one crop to another, or one genebank to another (Table 5.1). Within this category should be included accession descriptors which refer to the identification of germplasm within any one genebank, and collection descriptors referring to how, when and where the germplasm was originally collected.

Table 5.1 Recommended passport descriptors for accession and collection data. (From IBPGR, 1983b.)

1. ACCESSION DATA
   - 1.1 ACCESSION NUMBER
     This number serves as a unique identifier for accessions and is assigned by the curator when an accession is entered into his collection. Once assigned this number should never be reassigned to another accession in the collection. Even if an accession is lost, its assigned number is still not available for re-use. Letters should occur before the number to identify the genebank or national system (e.g. MG indicates an accession comes from the genebank at Bari, Italy. PI indicates an accession within the USDA system)
   - 1.2 DONOR NAME
     Name of institution or individual responsible for donating the germplasm
   - 1.3 DONOR IDENTIFICATION NUMBER
     Number assigned to accession by the donor
   - 1.4 OTHER NUMBERS ASSOCIATED WITH THE ACCESSION (other numbers can be added as 1.4.3 etc.)

     Any other identification number known to exist in other collections for this accession, e.g. USDA Plant Introduction number (*not* collection number, see 2.1)

     1.4.1 *Other number 1*
     1.4.2 *Other number 2*
   - 1.5 SCIENTIFIC NAME
     1.5.1 *Genus*
     1.5.2 *Species*
     1.5.3 *Subspecies*
     1.5.4 *Botanical variety*

continued

Table 5.1 continued

1.6 PEDIGREE/CULTIVAR/TYPE/NAME
Nomenclature and designations assigned to breeder's material

1.7 ACQUISITION DATE
The month and year in which the accession entered the collection, expressed numerically, e.g. June = 06, 1981 = 81
1.7.1 *Month*
1.7.2 *Year*

1.8 DATE OF LAST REGENERATION OR MULTIPLICATION
The month and year expressed numerically, e.g. October = 10, 1978 = 78
1.8.1 *Month*
1.8.2 *Year*

1.9 ACCESSION SIZE
Approximate number of seeds of accession in collection

1.10 NUMBER OF TIMES ACCESSION REGENERATED
Number of regenerations or multiplications since original collection

2. COLLECTION DATA

2.1 COLLECTOR'S NUMBER
Original number assigned by collector of the sample normally composed of the name or initials of the collector(s) followed by a number. This item is essential for identifying duplicates held in different collections and should always accompany sub-samples wherever they are sent.

2.2 COLLECTING INSTITUTE
Institute or person collecting/sponsoring the original sample

2.3 DATE OF COLLECTION OF ORIGINAL SAMPLE
Expressed numerically, e.g. March = 03, 1980 = 80
2.3.1 *Month*
2.3.2 *Year*

2.4 COUNTRY OF COLLECTION OR COUNTRY WHERE CULTIVAR/VARIETY BRED
Use the three letter abbreviations supported by the Statistical Office of the United Nations. Copies of these abbreviations are available from the IBPGR Secretariat and have been published in the FAO/IBPGR Plant Genetic Resources Newsletter number 49.

2.5 PROVINCE/STATE
Name of the administrative subdivision of the country in which the sample was collected

2.6 LOCATION OF COLLECTION SITE
Number of kilometres and direction from nearest town, village or map grid reference (e.g. TIMBUKTU7S means 7 km South of Timbuktu)

2.7 LATITUDE OF COLLECTION SITE
Degrees and minutes followed by N (north) or S (south), e.g. 1030S

2.8 LONGITUDE OF COLLECTION SITE
Degrees and minutes followed by E (east) or W (west), e.g. 7625W

2.9 ALTITUDE OF COLLECTION SITE
Elevation above sea level in metres

continued

2.10 COLLECTION SOURCE

1. Wild
2. Farm land
3. Farm store
4. Backyard
5. Village market
6. Commercial market
7. Institute
8. Other

2.11 STATUS OF SAMPLE

1. Wild
2. Weedy
3. Breeders line
4. Primitive cultivar (landrace)
5. Advanced cultivar (bred)
6. Other

2.12 LOCAL/VERNACULAR NAME
Name given by farmer to cultivar/landrace/weed

2.13 NUMBER OF PLANTS SAMPLED
Approximate number of plants collected in the field to produce this accession

2.14 PHOTOGRAPH
Was a photograph taken of the accession or environment at collection?
O No
+ Yes
2.14.1 *Photograph number*
If photo has been taken provide any identification number

2.15 HERBARIUM SPECIMEN
Was a herbarium specimen collected?
O No
+ Yes

2.16 OTHER NOTES FROM COLLECTOR
Collectors will record ecological information. For cultivated crops, cultivation practices such as irrigation, season of sowing, etc. will be recorded

---

The one most important descriptor for any genebank is that of **accession number**. Units of germplasm within a genebank should each be given a single, simple, different number falling sequentially, which will serve always to identify that germplasm. This will be a 'unique identifier'. The second most important descriptor may well be that of **collection number**, and should be a number given to germplasm when it is first collected in the field. This should be a number uniquely identifying germplasm to any individual collector or collecting institute. If the collector's name and number always accompany germplasm when it is distributed in any way, then it will serve as a means of avoiding otherwise undetectable and wasteful duplication of resources throughout the international community.

Genetic value of descriptors

The other three categories of descriptor will be used for information regarding specific traits and will be more or less understood. However, until recently, very little consideration has been given to the quality of descriptors in terms of their genetic information content. This is in fact hinted at in the IBPGR categories of descriptors given above. It is after all, the genetic aspect of any plant character which is of importance to plant breeding. This idea should lead us to seek the clearest possible definition of characters in the plant, and to define characters as the smallest possible heritable units. As an example, 'yield' is often used as a descriptor of evaluation. It is certainly an interesting and

Table 5.2 Seed colour in peas demonstrating that descriptors can be used on the different genetic and botanic levels. (From Blixt and Williams, 1982.)

Level	*Descriptor*	*No.*	*Descriptor states*
A. *Pisum sativum*			
Monogenic	M		Marbled Not marbled
	Gri		With grey median stripe Without grey median stripe
	I		Cotyledons yellow Cotyledons green
	Gla	8	Green colours bluish Green colours not bluish
Oligogenic	Seed colour testa		Marbled with grey stripe Marbled without grey stripe Not marbled with grey stripe Not marbled without grey stripe
	Seed colour cotyledons	7	Cotyledons yellow Cotyledons green Cotyledons bluish green
Polygenic	Seed colour	12	Testa marbled, with stripe, yellow Testa marbled, with stripe, green Testa marbled, with stripe, bluish green Testa marbled, without stripe, yellow Testa marbled, without stripe, green Testa marbled, without stripe, bluish green Testa not marbled, with stripe, yellow Testa not marbled, with stripe, green Testa not marbled, with stripe, bluish green Testa not marbled, without stripe, yellow Testa not marbled, without stripe, green Testa not marbled, without stripe, bluish green
B. *P. arvense*			
Monogenic		26	54
Oligogenic		8	90
Polygenic		2	640
		1	about 1 000 000
		1	about 4 000 000

important one describing the result of complex interactions. However, it cannot really be said to be a single character, because of its complexity. What is required therefore is the assignation of descriptors to the actual genes which, if known, control the components of yield and the expression of which results in the yield.

For many crops and many characters, the results of genetic analysis are just not available. This may well be the state of affairs for many years to come. Providing characters are distinct and recognizable they can be represented by clear and distinct descriptor states, expressed for example as a numerical scale of integers.

Looking to the future, though, the aim should be to make descriptors more 'genetically accountable', where the descriptor states will be the alleles of a known gene. At the Nordic Genebank (NGB) and the Weibullsholm Plant Breeding Institute (WBP) in Sweden, breeders have gone a long way towards this goal (Blixt and Williams, 1982; Blixt, 1984). They have been assisted by the wealth of genetic knowledge accumulated for the genus *Pisum* (Blixt, 1978). The pea descriptors can be managed at different botanic and genetic levels (Table 5.2) designated as polygenic, oligogenic and monogenic. One practical difference in descriptors at these levels lies in their number of descriptor states (Blixt, 1982). At the polygenic level the pea descriptor 'seed colour' results in very large numbers of descriptor states. However, if at the oligogenic level, one component of seed colour, i.e. cotyledon colour, is taken, eight gene descriptors each with roughly 11 allele descriptor states can be recorded. On the monogenic level, 26 gene descriptors account for variation in seed colour, with about two descriptor states each. Genetic representation therefore requires more descriptors, but these can be much more precisely defined, and do much more to preserve valuable information about important plant material.

Referring once again to the quality of information inherent in any descriptor, this may secondarily depend upon the way that descriptor states are organized. If descriptors which represent genes are being dealt with, then deciding upon descriptor states presents few problems because they will be made up of recognized gene symbols. On the other hand for passport descriptors and other characterization and evaluation descriptors where information about genetic control is lacking, some form of coding for descriptor states may be useful and necessary (Table 5.3). The use of a letter code for instance (alphanumeric) will result in space saving in the data preparation documents. Properly selected codes might prove less cumbersome for a scientist to manipulate than the full-length representations, providing the codes are explicit enough to make the need for constant reference to a 'code dictionary' unnecessary. The descriptor 'country of origin' exemplifies descriptors for which letter coding is appropriate. For example, having once encountered the descriptor state code 'PER', it would not be difficult to remember that this code stands for the country Peru. Numeric codes can provide several benefits also. Numbers greatly facilitate statistical analysis such as frequency distributions, and measurements of central tendency for instance. In particular, numeric codes are not language dependent, so a data set utilizing them might be more internationally acceptable. The code value '1' for example, could be defined in a code dictionary as BLANCO just as easily as WHITE. Where it is necessary to code for such characters as plant height, or seed weight for example, then ranges of values can be represented by a 1–9 scale.

Table 5.3 Some examples of coded descriptors for *Capsicum*. (From IBPGR, 1983.)

Descriptor	*Descriptor code*	*Descriptor state*
Plant growth habit	3	Prostrate
	5	Compact
	7	Erect
Fruit colour in mature stage	1	Green
	2	Yellow
	3	Orange
	4	Red
	5	Purple
	6	Brown
	7	Black
	8	Other
Fruit Length	1	Very short – <1cm
	3	Short – around 5 cm
	5	Medium – around 10cm
	7	Long – around 15 cm
	9	Very long – >25cm
Fruit persistence	0	Deciduous
	+	Persistent
Susceptibility to drought (based on a 1–9 scale)	3	Low susceptibility
	5	Medium susceptibility
	7	High susceptibility
Susceptibility to *Alternaria solani* (based on a 1–9 scale)	3	Low susceptibility
	5	Medium susceptibility
	7	High susceptibility

The requirements of a computerized documentation system

As a genebank preserves and provides genetic material for a multitude of purposes, in order to meet the demands of both current and future users, any data management system should be able to accept all data relevant to the genetic material. It should be able to deal with varying frequencies of data capture, optimizing the storage capabilities of the system under such conditions. There is a need for 'open' descriptor lists, so that new information can be introduced at any time. Functions like adding, removing and redefining descriptors will be helpful in this respect.

The communication between user and system should be simple enough to permit the user to be a botanist or plant breeder say, rather than a computer scientist. An interactive system probably holds out the greatest hope of friendliness towards the user, although 'batch' input of data would be a necessary additional alternative if at the initiation of a computerised data management system, substantial amounts of data already existed in computer-readable form. In an interactive environment, the system may continually prompt the user for information throughout operation. Alternatively the system may provide a problem-oriented, rather than computer language, so that the user can easily formulate questions and instructions to be acted-upon perhaps with the aid of a screen 'menu'.

Being able to process data files produced by other systems, as well as being able to produce data files acceptable to other systems will be of considerable benefit. It will form the basis for information exchange between different installations or genebanks. In addition to this, exchange of information in more visually acceptable printed form will be enhanced by the availability of some form of report generator. A checklist of some other features of importance is given in Table 5.4.

Table 5.4 Checklist of additional useful features for data management systems.

1. Good data entry mechanism. Preferably screen driven input with validation.
2. Friendly user interface – concise meaningful commands, good error messages.
3. Validation on entered data perhaps including check digits on accession number.
4. Fieldnames to be long enough to be recognisable, preferably to have long and short forms.
5. Ability to reflect descriptions of coded descriptor states.
6. Easy updating of existing fields including replacing missing values.
7. Allows arithmetic on fields, e.g. field = field − 100.
8. Good quality output, variable formats.
9. Ability to sort on output.
10. Allows upper and lower case data.
11. Ability to interface with other programmes.

Models for data management

A simple system using a single data file will be of somewhat limited use particularly for very large quantities of data, but it would be better than nothing! It could be a useful first move towards a computerized system, provided the structure of the file was fairly universally acceptable. At the other extreme is a full-bodied 'database management system'. This can be seen as a collection of files required for individual applications and the database management system is made up of programmes used to create, maintain and reference these files. These programmes show independence of the data; they do not need to be concerned about the physical location of the data they process. The beauty of such systems lies in their flexibility, and particularly in their data independence which should allow single data files referred to in the first model to be incorporated into the database.

Two basic types of database management system can be identified, namely hierarchical and relational. A hierarchical system tends to be extremely complex because of the 'relationships' or 'connections' which are set up between the data elements. A relational database management system is much simpler and probably more appropriate. Data are represented as they are, and the 'relations' between data elements can be considered in the form of two dimensional tables. Each row across the table is called a record (referring to an accession, seed lot, plant etc.) and each column is called a field of the record (containing individual descriptors). Information contained in any two or more separate tables (for example, one for passport, and the other for genebank maintenance) can then be related or combined if there are common fields (or descriptors) in those tables. This will often act as a means of creating 'new' files within the database for further manipulation.

In simple terms, what these systems will do in varying degrees is to model the good old-fashioned filing cabinet, but with links between the files, so that one can have for instance a file of germplasm collection details linked to other files for different evaluation trials, where the link will probably be made through a common descriptor or key such as accession number. Most of these systems are aimed at the end user, and are driven by 'menus'. More advanced systems may come with command languages which permit the tailoring of the package to particular applications.

Data preparation

One major pitfall which traps many people hastily rushing to computerize their data, is the bypassing of adequate and reliable data preparation procedures. Regardless of the computerized data management system to be used, other management techniques need to be applied first of all. Starting with any original 'raw' data source, it is necessary to pass, one step at a time, through a set of procedures for describing, extracting and transcribing the data. The resulting machine-readable data set should be adequately structured and sufficiently documented to allow an analyst who is unfamiliar with the subject matter to process the data with a variety of computerized information systems. Continuity of personnel within any institute or genebank can never be guaranteed. None of us lives forever! Some of us may even get promoted. If such change happens suddenly and unpredictably then the question to ask is 'can someone else take over the task of managing the genebank data to ensure that work carries on?' Efficient data preparation and documentation of procedures will solve such problems.

The sources from which data are to be extracted, be they fieldbooks, laboratory notebooks, collection forms or evaluation trials sheets, need therefore to be documented. All descriptors of interest need to be listed thoroughly and to be defined, so that further recording or measurement will be standardized (Table 5.5). Where descriptor codes are employed, it will be useful to construct a code dictionary listing the acceptable values for each code (Table 5.3, p. 76). Reference should be made to the physical representation of the data, in terms of whether or not values will be characters or numeric, and whether numeric values involve decimal places for instance. The final format of the prepared data on paper will depend to some extent upon the mode of data entry into a computer. Batch mode would tend to require a final format in the form of a data table constructed in such a way as to make entry of data into the computer as easy and reliable as possible. On the other hand interactive screen entry using a visual display unit (VDU) would be better served by the use of a data sheet designed to look like the 'form' appearing on the VDU. Here a single data sheet per accession would be needed (see Fig. 3.2, p. 48). With some thought as to design, this sheet or form could easily be used for other purposes earlier on in the data capture process. In other words a single form could be used at the time of data collection (even in the field) which would be suitable for direct data entry into the computer at a later date. One major advantage to this is that no errors would be introduced during the unnecessary processes of transcription.

Table 5.5 Some examples of descriptor definitions for wheat. (From IBPGR, 1981.)

Donor name	The institution or person responsible for donating the germplasm to the collection
Date of collection of original sample	Expressed as day/month/year, e.g. 20 October 1981 is recorded as 201081
Spike density	A visual measure of density of a spike measured on a 1–9 scale, where 1 is very lax and 9 is very dense (N.B. spike density is not the same as spike shape)
Plant height	Height of plant at maturity, measured in centimetres from ground to top of spike, excluding awns.
Days to flower	Counted as days from sowing to 50% of plants in flower. However, when planting in dry soils in dryland areas it is counted from the first day of rainfall or irrigation which is sufficient for germination.
Percentage protein content	Measured as percentage of dry weight (seed mositure equal to or less than 12 percent), indicating the conversion factor used as either N × 6.25 or N × 5.6

Table 5.6 Procedures for efficient data preparation.

1. Formulate descriptor list
2. Define descriptors
3. Revise or check descriptor list
4. Define physical representation of data i.e. descriptor states, descriptor coding
5. Produce code dictionary
6. Design data table and/or final data record sheet as permanent record and for data entry into computer

The procedures for data preparation need careful consideration, as future universal use of an existing system for data management will depend upon them. It is worth listing the steps which are involved (Table 5.6).

Examples of systems in use

Many institutes involved in genetic resources work, and many genebanks also, are actively utilizing computerized systems for data management purposes. This encompasses a wide range of software programmes and packages. In the early days of IBPGR it was generally thought possible to develop a computer system specifically for genetic resources data handling, which could be used throughout the world. Some degree of progress was made towards this aim during the 1970s, when a system called TAXIR (Taxonomic Information Retrieval) and subsequently EXIR (Executive Information Retrieval) were developed. These are indeed used in a few genebanks in various places (Izmir, Turkey and Bari, Italy), but more widespread acceptance of the system has been restricted because of difficulties in portability. It has also been overtaken by fully-supported, commercial database management software available for a range of different computers, micro-, mini- and main-frame. Nowadays, the idea that everybody working with germplasm needs to use one and the same software package is no longer considered logical. Access to computing hardware is the first consideration. Once this has been thought about, then a choice of software is likely to open up.

One of the front-runners in the management of genetic resources data is certainly the Nordic Genebank (and Weibullsholm Plant Breeding Institute) where the information process is handled by a suite of programmes written in BASIC and run on a Wang 2200 minicomputer. At the National Vegetable Research Station, UK a computer system called INFO (DORIC) is operated on the station's VAX 11/750 minicomputer. A computer terminal and printer within the genebank permits rapid access, manipulation and presentation of stored information. At the University of Birmingham, small databases (describing about 5000 accessions) are easily maintained using a 16-bit Heath Z100 microcomputer and dBASE-II, a relational database management system developed by Ashton-Tate. At the other extreme, a large mainframe computer (IBM) is needed to support GRIP (Genetic Resources Information Programme) for the USDA germplasm collections. This utilizes a data management and statistical analysis package called SAS (Statistical Analysis System).

The truth is that now, and more so in the future, we will be bombarded with newer and better software and hardware capable of fulfilling our needs for efficient genetic resources data management. The choice is ours to make. The criteria that we set for ourselves and our equipment need to be determined with care. Having done this, we can then feel safe to take the plunge into these rapidly moving waters of technological innovation.

6
Utilization

The exploitation of genetic diversity for crop improvement should be the ultimate objective of genetic resources exploration and conservation. While it is important to ensure that plant genetic resources are adequately safeguarded for future generations, they cannot just be placed in a genebank and forgotten about. The vital stages of evaluation and incorporation of valuable characters such as disease resistance or tolerance of environmental stress factors into new varieties, are the justification of genetic resources activities.

The immediate value of genetic resources depends to a considerable extent upon the ease with which the breeder can utilize them. For instance, the primitive cultivars of crops are more closely genetically allied to modern varieties belonging to the same biological species than are the wild species. As a consequence the transfer of genetic characteristics is generally much easier from the primitive cultivars. Most plant breeders prefer to cross plants that belong to the same biological species, and which generally produce fertile hybrids and present little or no hindrance to genetic recombination. Under some circumstances, however, breeding progress can only be made by resort to wide crosses, which inevitably involves considerably more effort. Plant breeders are increasingly becoming aware of the importance of wild relatives of crop plants, and have long been urged by crop plant evolutionists to exploit this source of germplasm. Zohary *et al.* (1969) have pointed out the importance of the diploid wheats which have remained unexploited by breeders. As Harlan (1976) has stressed, the need for genetic variation and sources of resistance in the future will lead to the exploitation of all available sources of diversity.

In the meantime, the most rapid advances in plant breeding will be made when closely related parents are utilized. In order to utilize efficiently more distantly related species through inter-specific hybridization, it is convenient to have information available on the relationship between the species involved. Crop plant evolution studies based on experimental hybridization can determine the relationship of one species to another. Scientists working with primitive cultivars and wild species, and particularly those with interests in crop evolution can undertake so-called pre-breeding studies, in order to provide breeders with valuable information. Plant breeders generally have little time for these more basic studies, because their primary objective is to develop new varieties. Nevertheless, data from pre-breeding studies do allow the plant breeder to evaluate rationally the chances of success of incorporation of genetic material through wide crosses before the costly process of breeding actually commences.

Although originally devised as a method of classifying cultivated plants, the gene pool concept of Harlan and de Wet (1971) provides a useful scheme for the presentation of data from pre-breeding studies. Species which belong to the same gene pool as the cultivated species can be utilized relatively easily in breeding, but the use of species from other gene pools presents some difficulties, or hybrids are sterile or lethal (Fig. 6.1). Manipulation of species in wide crosses through bridging species which are cross-compatible with both is one way in which genes may be transferred from one species to another. In this text it is not our intention to include a discussion of crossability barriers. There are several pre-fertilization and post-fertilization isolation mechanisms which have been recognised between plant species. These are described in detail by Stace (1980), and are applicable to cultivated species and their wild relatives.

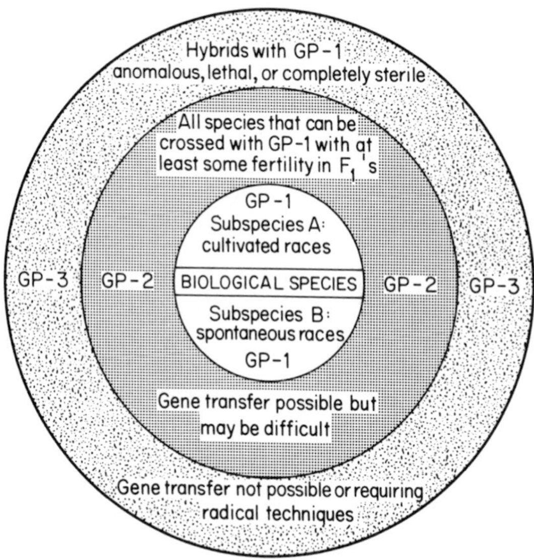

Fig. 6.1 Schematic diagram of primary gene pool (GP-1), secondary gene pool (GP-2) and tertiary gene pool (GP-3). (From Harlan and de Wet, 1971.)

To give an indication of the value of genetic resources in plant breeding, examples are given below of utilization of genetic resources in the eight crops we have chosen to cover; furthermore the location and scope of important germplasm collections are given in the Appendix.

Wheat

Wheat breeders can draw from a large gene pool notable for its diversity. There are at least 23 species of *Triticum* with ploidy levels of 14, 28 and 42 chromosomes. Emphasizing Harlan and de Wet's gene pool concept, the wheat primary gene pools include types that cross readily with one another;

landraces of the cultivated forms at the different ploidy levels will come in here. A very large secondary gene pool of about 35 species includes all species of *Aegilops, Secale, Haynaldia* and some of *Agropyron*. Gene transfer is possible with members of the primary gene pools, but special techniques may be needed.

Several very large collections of wheat germplasm exist; it is however very difficult to judge to what extent these represent an adequate proportion of the world total of wheat diversity.

Because of the polyploid nature of the major cultivated wheat forms, the accumulation and preservation of many and diverse genes has been facilitated during the history of cultivation. The repetition of genetic information throughout the sets of chromosomes of wheat and its near relatives has served as a buffer system that has allowed an enormous number of natural mutations to occur, survive, and add to the genetic diversity of wheat known to be available to the plant breeder. Examples of useful and utilizable genes are rather too numerous to list, but we can be certain that even if the major world wheat germplasm collections are not fully representative of the natural diversity that has at some time existed, it will be enough for the plant breeders to go on with for many years.

The other point of interest which affects the intraspecific utilization of wheat germplasm and is again associated with polyploids, is that of chromosome manipulation. Whole chromosome manipulation is facilitated by the duplication of genes permitting the loss of chromosomes without disastrous consequences to a plant's phenotype. In wheats, the use of aneuploids has been most extensive. A particular example of their use has been the introduction of chromosome 5A of *Triticum spelta* into winter wheats to convert them to spring wheats. More generally, methods using reciprocal monosomics afford the possiblity of screening for the existence of specific chromosomal variation among some of the most important wheat germplasm in the world where monosomic series exist (Law *et al.*, 1981).

Maize

No better example exists of the need to be able to broaden the genetic base of any crop plant, than that of maize in the early 1970s. For years before this, hybrid cultivars had proved enormously successful. In maize, which is wind pollinated, these hybrids could be easily produced by cross fertilization induced simply by cutting off the male inflorescences of all the plants intended as the female parent, while allowing the intended male plants to release their pollen normally. Even so, the process of detasseling proved laborious and costly in manpower, so that the production of hybrid varieties was greatly enhanced by the introduction of the important genetic tool for controlling fertilization, namely cytoplasmic male sterility. The development of a source of cytoplasm conferring male sterility certainly gave a boost to the production of hybrid maize, not only in the USA but throughout the world, with new hybrids both increasing yield levels in traditional maize-growing areas and extending the range of the crop into previously marginal ones. This success however, has to be set against the disastrous attack in the USA of a previously unimportant disease, southern leaf blight (caused by *Helminthosporium maydis*), which occurred in 1970 on all the hybrids based on this one cytoplasm, Texas-T. The reason was that this single, genetically limited cytoplasm carried the genes for disease susceptibility as well as for male sterility. This dramatically demonstrates the dangers of relying upon a single type of cultivar for large areas of production, and the need for broad genetic resources to be available for breeding. Fortunately in maize, resources were available from which to develop other forms of cytoplasmic male sterility not associated with disease susceptibility, although these are not proving as satisfactory as the original T-cytoplasm, heralding a return to mechanical detasseling for hybrid production.

In practice, the genetic base for maize has been and can be broadened through the development of various gene pools. These may include either indigenous or more exotic sources of genetic variation. Certainly, if the genetic diversity between source populations used to produce hybrid varieties is maximized, this will yield the best results (Williams, 1981). The highly successful early hybrids of maize resulted because of the diversity inherent in the introgression of Southern Dent and North Eastern Flint genotypes. The same principle has emerged from more recent hybrid maize programmes in the tropics. Inbred lines derived from diverse genetic stocks have generated hybrids with outstanding agronomic performances, which have quickly replaced the existing non-hybrid varieties adapted to tropical regions (Wellhausen, 1978).

There are often good reasons for attempting to utilize genes known to occur in more exotic sources of germplasm, by performing wide-crossing. This may be because a desired trait cannot be found within a given crop or species, but may be present in a more distant or unrelated species. Such is the case with maize. At CIMMYT for instance, scientists have been concentrating on wide crosses involving maize and sorghum and in particular, maize and *Tripsacum* (Peloquin, 1981). Current emphasis at CIMMYT in the 'Maize Wide Cross Program' is on maize and *Tripsacum* crosses, and their resulting progeny. *Tripsacum* is known to possess broad adaptation, stress tolerance to water-logging and drought, and resistance to foliar diseases and insect pests. Illustrating the extent to which such wide-crossing programmes utilize genetic resources, over 130 F_1 hybrids are maintained, and these have involved five different maize gene pools in combination with 17 different collections, representing six taxa of *Tripsacum*.

The manipulations necessary to bring such alien germplasm into plant improvement programmes are often complex and delicate. Harlan and de Wet (1977) have pioneered one technique that is now being used for maize. With this, sequential backcrosses of the F_1 with maize are made, together with the maintenance of the F_1 chromosome number. After several generations, the backcross plants with the normally expected chromosome number are selected and backcrossed again. Plants are then selected for near-normal maize chromosome number and also for phenotypic traits. They are then recombined to form a population used for screening for desired characters.

Other more novel techniques are also being examined, such as the generation of somaclonal variation, embryo rescue and micromanipulation.

These techniques are referred to in more detail in the final chapter.

One of the most recent genetic resources to receive publicity is the 'billion dollar root' belonging to the species *Zea diploperennis*. This may hold out enormous potential for future maize breeding and cultivation. As its name implies it is a perennial, where genes conferring longevity may easily be transferred to *Zea mays*. Although a number of experts assert that perennial cereals do not yield as much as annuals, hybridization with *Z. diploperennis* could serve to extend the cultivation range of the maize crop by as much as 10% because of the adaptation of the wild species to fairly extreme mountainous terrain. As a further bonus, it seems that this species also possesses immunity or tolerance to at least four major virus diseases – truly a resource amongst resources?

Rice

Since 1972 the International Rice Research Institute (IRRI) in the Philippines has been working along with seven other countries in Asia to conserve rice germplasm. Well over 63 000 seed samples mainly of rice cultivars are maintained and also channelled into IRRI's Genetic Evaluation and Utilization programme (Fig. 6.2).

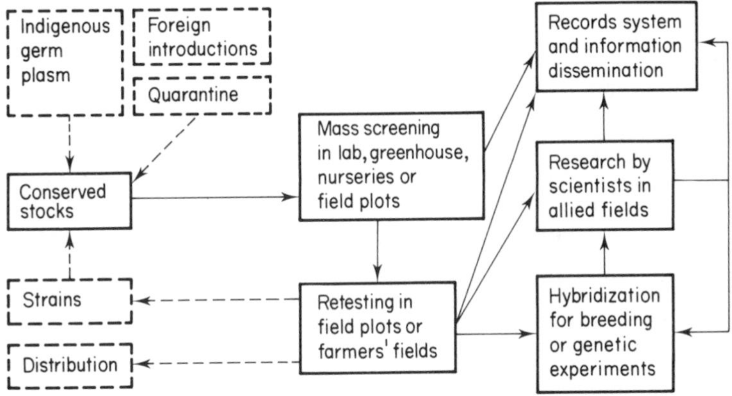

Fig. 6.2 IRRI rice evaluation programme. (From Chang, 1976.) Flow chart outlining the necessary steps in the evaluation and utilization of rice germplasm in relation to conservation and dissemination. Operations indicated by solid lines and arrows are directly related to evaluation, utilization, and information storage and retrieval. Dotted lines and arrows indicate indirect activities.

The semi-dwarf characteristics conferring wide adaptiveness and high yield potential which were originally developed in China and at IRRI using the 'Dee-geo-woo-gen' and other sources, are now widespread amongst modern rice cultivars. This means that the genes that many rice breeders are searching for are primarily those needed to eliminate defective features such as susceptibility to certain diseases or insects or lack of specific adaptation to adverse environments. Limiting factors found in the environments of low-yielding rice areas may be due to lack of irrigation, or excessive water, soils where edaphic factors are in deficit or excess, low temperatures at high altitudes or latitudes, or high temperatures during the reproductive stages of the growth period.

Several wild species of rice hold potentially useful genetic diversity, and strains of annual forms have been used in breeding programmes (Chang *et al.*, 1977). For both *Oryza sativa* and *O. glaberrima* it has been reported that sufficient genetic diversity still exists (Chang *et al.*, 1975). In Bangladesh for instance, many deep water cultivars have outstanding ability for internode elongation within four weeks of seeding, and other local rices are highly tolerant of flood submergence. Bangladesh strains have also provided the xa_5 and xa_7 genes for resistance to bacterial blight (Sidhu *et al.*, 1978). *Javanica* varieties now grown in the Malagasy Republic, originating in Indonesia, are reported to have tolerance to cool temperatures, while 'Silewah', an *indica* variety collected in Sumatra, is apparently highly salt tolerant (Satake and Toriyama, 1979). Resistance genes for bacterial blight and tungro virus can also be found in a high number of locally adapted cultivars of Indonesian origin. The genes Bph 3 and bph 4 are thought to originate in Sri Lanka (Khush, 1977). These confer broad resistance to the brown planthopper of which several biotypes have been identified.

Barley

Major barley germplasm collections continue to function as important sources of useful genes, particularly in terms of natural mutants conferring high amino acid levels and high total protein contents. A survey of over 4000 barley accessions, mainly landraces held in the Gatersleben genebank in the German Democratic Republic, resulted in the selection of several lines and cultivars with high lysine and total protein (Lehmann *et al.*, 1976). In the USDA world barley collection maintained at the Beltsville genebank, 113 cultivars were screened for protein content, amino acid composition and kernel weight (Pomeranz, 1976). It was found that protein content was highest in naked 6-rowed spring cultivars and lowest in hulled 6-rowed winter varieties. The latter also showed the highest lysine plus threonine percentage of protein. Large differences among these cultivars indicated the presence of a rich source of variation in the collection of potential use to plant breeders.

The discovery of the high protein and high lysine barley line 'Hi-proly' by Munck and co-workers (1970) may serve as another example of the genetic potential which barley germplasm collections hold for breeders. The Hi-proly cultivar CI 3947 is an Ethiopian line from the USDA barley collection, and is considered to provide more essential amino acids than any other commercial grain cultivar including the best maize mutants (Pomeranz, 1973), but barring the barley mutant Riso 1508 which has 45% more than average lysine content.

For a number of years now, plant collectors have recognized that Ethiopia is of utmost importance when considering diversity in barley. Giessen, Hoffman and Schottenloher in 1956 reported an analysis of collections made in Ethiopia in 1937–38. They identified variation in several characteristics considered of importance to plant breeders: tillering, straw length, kernel size, disease resistance (mildew and dwarf rust) and high protein content. Qualset and Suneson (1966) later constructed a breeding population which was developed to incorporate many of these characters of Ethiopian barley into adapted genotypes.

Further recent collections of barley from the Yemen Arab Republic made by a team led by Ayad (Ayad *et al.*, 1980) have been screened for disease resistance. They have been found to vary in reaction to loose smut (*Ustilago nuda* var. *hordei*) (Damania and Porceddu, 1982) and also to possess a 'non-heading' character in which the spike fails to emerge from the flag leaf even at maturity. It is suggested that this may be a character of use to the plant breeder since it not only protects the spike from fungal spores but also restricts plant height which may prove useful in the prevention of lodging (Damania *et al.*, 1985).

Collections made in Israel of the wild *spontaneum* type are currently proving to be useful to Moseman *et al.* (1983) for developing improved barley cultivars, as they have found widespread sources of resistance to powdery mildew (*Erysiphe graminis*).

Potato

The potato represents one of the best examples of the utilization of wild species and primitive cultivars in plant breeding. Why should this be so? Firstly, the early importance of the potato as a staple food in Europe and North America has undoubtedly contributed to a long history of breeding efforts. The narrow genetic base of the potato was recognized well over 100 years ago after the late blight epidemics of the 1840s, caused by the fungus *Phytophthora infestans*, devastated the potato crop, particularly in Ireland. Secondly, early attempts were made to survey and utilize genetic variability in the potato. From 1925, several Russian expeditions collected potatoes in Central and South America, and laid the foundations for our understanding of the polyploid nature of the potato and its wild relatives. Thirdly, the pattern of genetic variability in the potato is very different to other major crops such as wheat and barley. In these crops, the wild species are restricted in number and variability, although considerable variation does occur in the cultivated gene pool. In contrast, there are almost 200 wild species of potatoes, as well as several thousand primitive cultivars, and most of the genetic characters are available to the plant breeder. The effect of polyploidy and partial sterility make it difficult to transfer genes from certain species to *Solanum tuberosum*, but nevertheless there has been considerable success in utilizing wild species in wide crosses. Ploidy differences have been overcome through the functioning of unreduced or 2n gametes, i.e. gametes with the somatic chromosome number, so that the diploid gene pool is made available more readily to the breeder. Over 70% of the tuber-bearing Solanums are diploid, and the phenomenon of 2n gametes is widespread in both wild and primitive cultivated diploid species (Quinn *et al.*, 1974; Camadro and Peloquin, 1980).

The crossability groups of potatoes are well defined. In difficult crosses, it has been possible to utilize bridging species such as *S. acaule*, a high altitude frost-resistant species from Peru and Bolivia, in order to effect gene transfer.

Certain allotetraploid species such as those belonging to the Mexican Series Longipedicellata behave as diploids, and do not cross easily with *S. tuberosum*, although it appears that the crossability barriers are post-fertilization. In these and other wide crosses where post-fertilization breakdown of the endosperm, and embryo abortion are frequent, the techniques of embryo culture offer considerable promise in making these sources of genetic diversity available to breeders. In collaborative research between the International Potato Center, the Agricultural University, Wageningen in the Netherlands, and the University of Birmingham, successful transfer of resistance genes to potato leafroll virus (PLRV) has been achieved from distantly related species such as *S. etuberosum* and *S. pinnatisectum* through the culturing of hybrid embryos. The extensive development of pre-breeding studies in the tuber-bearing Solanums is another reason why wild species have been utilized so widely in breeding.

The decision to incorporate wild species into a breeding programme is an important one because of the problems associated with their use, such as undesirable characteristics which are difficult to eliminate through a protracted programme of backcrossing. In this respect the World Potato Collection maintained by CIP in Peru represents an invaluable resource. Extensive screening of this large collection of several thousand primitive cultivars has led to the identification of resistance to many of the major pests and diseases. Such sources of genetic diversity can be used with comparative ease, because the primitive cultivars are closely related to *S. tuberosum*. Nevertheless, it has been necessary to turn to the wild species for some valuable genes, and many modern varieties have wild species in their pedigrees. Important sources of resistance to diseases, pests and frost injury are shown in Table 6.1.

New varieties are needed for several reasons. Often they are introduced because their yields are higher than current varieties, or they have better quality or resistance to pests and diseases. One important success story in terms of germplasm utilization is the transfer of resistance to the bacterial wilt pathogen, *Pseudomonas solanacearum*, from the primitive diploid cultigen, *S. phureja* to *S. tuberosum*. Resistance to bacterial wilt was first found by Thurston and Lozano (1968) in the Colección Central Colombiana, in Colombia. Although occurring in diploids, this resistance has now been successfully incorporated into several potato varieties, and the functioning of 2n gametes in the diploids has facilitated the breeding process.

Perhaps the earliest utilization of a wild species was the transfer of resistance to late blight from the wild Mexican hexaploid species, *S. demissum*, in which resistance had been demonstrated as early as 1909 by Sir Redcliffe Salaman in the United Kingdom. This major gene source of resistance was important for several decades. In recent years, there has been a move towards sources of horizontal or field resistance to late blight from *S. tuberosum* ssp. *andigena*. The potato cyst nematodes, *Globodera rostochiensis* and *G. pallida* are major pests in many countries. Resistant cultivars have been developed with resistance derived from a wild diploid species, *S. vernei*, from north west Argentina.

The availability of a large and diverse gene pool, and its successful utilization

Table 6.1 Important sources of pest and disease resistance and frost tolerance amongst the tuber-bearing Solanums.

Pest, disease or stress	*Source of resistance*
Late blight (*Phytophthora infestans*)	*S. demissum* (2n = 72) *S. tuberosum* ssp. *andigena* (2n = 48)
Early blight (*Alternaria solani*)	*S. chacoense* (2n = 24)
Bacterial wilt (*Pseudomonas solanacearum*)	*S. phureja* (2n = 24) *S. sparsipilum* (2n = 24)
Potato virus X	*S. acaule* (2n = 48)
Potato virus Y	*S. stoloniferum* (2n = 48)
Potato leafroll virus	*S. etuberosum* (2n = 24)
Potato cyst nematode (*Globodera* spp.)	*S. vernei* (2n = 24)
Root knot nematode (*Meloidogyne* spp.)	*S. microdontum* (2n = 24)
Colorado beetle (*Leptinotarsa decemlineata*)	*S. commersonii* (2n = 24)
Frost tolerance	*S. acaule* (2n = 48)

in potato breeding contrasts with the situation in many other crops. Undoubtedly the economic importance of the potato crop in the developed countries has been a catalyst for these efforts. Fortunately, international cooperative efforts are now aimed at producing potatoes for the small-scale farmer in developing countries where the incorporation of genetic resistance to pests and diseases will be necessary because of the inability of many farmers to use chemical control.

Breeding efforts have been matched by those in genetic conservation. In addition to the World Potato Collection of primitive cultivars, the International Potato Center also has an important collection of wild species, many of which are new to science. Important collections of wild species have been established for many years now in Scotland at the Commonwealth Potato Collection, in the United States at the USDA IR-1 Collection at Sturgeon Bay in Wisconsin, and at the Dutch-German Potato Genebank at Braunschweig in West Germany. The interrelation between breeding and pre-breeding has been a major factor in the success of breeding programmes in these centres.

Cassava

Interspecific hybridization has played an important role in the breeding of new cassava varieties, especially in Africa and India (Martin, 1976). However, relatively few wild species have been used in crosses with *Manihot esculenta*, including *M. glaziovii*, *M. saxicola*, *M. melanobasis*, *M. catingae* and *M. dichotoma*. The introduction of resistance to cassava mosaic disease has been achieved through crosses between *M. esculenta* and *M. glaziovii*. In many interspecific crosses, F_1 hybrids are fairly fertile, and even when hybrids are partially sterile it is possible to obtain fertile derivatives by backcrossing. The barriers to hybridization appear to be minimal.

Within *M. esculenta* there is great variability for characters such as root yield, harvest index, root dry matter content, HCN content and root perishability after harvest. In addition resistances to some of the major diseases and insects such as cassava bacterial blight (*Xanthomonas manihotis*), *Cercospora* leaf spot, *Phoma* leaf spot, and thrips have been found within the cultigen (Kawano, 1980). A germplasm collection of international importance has been established at CIAT. Nevertheless, cassava breeders need access to other *Manihot* species, and undoubtedly species such as those described above offer a great untapped source of new material for the plant breeder. Regretably, few of the species are maintained in living collections, although considerable information on the diversity in *M. esculenta* is now available following the publication of an important monograph by Rogers and Fleming (1973).

Sweet potato

Hybridization has largely been restricted to crosses within hexaploid *Ipomoea batatas*, because all but one of the wild species are either diploid or tetraploid species, and their utilization in sweet potato improvement consequently presents difficulties. According to Jones (1980), variation within *I. batatas* is extensive, and breeders have only just begun to utilize available germplasm.

A search of the literature indicates that primitive clones with resistance to several diseases including blackspot (*Ceratocystis fimbriata*) and soil rot (*Streptomyces ipomoea*) can be found extensively in South America. Plants collected in the Caribbean with a small growth form have been utilized in breeding programmes in New Zealand, because of high root yields, and because such forms were useful for mechanical harvesting.

Undoubtedly the variability in sweet potato varieties found now in the United States is much less than reported earlier this century. Sweet potato improvement for the tropics is still at a low level of development because of the low priority afforded the crop until recently. A germplasm collection is maintained at the Asian Vegetable Research and Development Center (AVRDC) in Taiwan.

Grain legumes

In the legumes, in contrast to many other plants, isolating mechanisms between species tend to be effectively developed, and individual species may well be completely isolated genetically from their closest relatives (Smartt, 1981). An understanding of the potential of gene exchange between cultivated species and wild species is therefore of great importance. Between some pairs of species isolating mechanisms may not be developed to the exclusion of all gene exchange, as the example of the common bean, *Phaseolus vulgaris* and some of its relatives illustrates. In contrast, *P. lunatus* and *P. acutifolius* are incapable of gene exchange with other related biological species. The application of the Harlan and de Wet gene pool concept is particularly valid in the grain legumes, and has been applied successfully in *Phaseolus* and *Vigna* (Smartt, 1981), *Lens* (Ladizinsky, 1979) and *Cicer* (Ladizinsky and Adler, 1976), not only for the identification of wild progenitors of cultivated species, but for the definition of close relatives for breeding purposes.

Many legume species are isolated by the presence of chromosome rearrangements which prevent pairing at meiosis. In the case of *Vicia* on the other hand, as well as in *Lathyrus*, it is apparent that pre-zygotic barriers operate. Although the utilization of the wild gene pool is hampered by such problems, it is clear that exploitation of wild species is becoming a common practice. In lentils, for example, the wild gene pool including *Lens orientalis* and *L. nigricans* can be exploited almost equally and their variation easily utilized in breeding (Ladizinsky, 1979). Furthermore, these wild relatives still possess important variation that no longer exists in their cultivated counterparts. The wild gene pool of lentil contains a remarkable number of disease resistance genes which are quite rare in the cultigen, *L. culinaris*. In the case of the common bean, *P. vulgaris*, however, considerable variation has accumulated in this species so that material from the same gene pool can be utilized in breeding.

The improvement of nutritional quality is one of the most important pulse breeding goals. The improvement of protein quality is particularly important in the developing world where young children are subject to protein deficiency in the diet. The gene pools of *P. coccineus* and *P. vulgaris* have extensive protein polymorphisms which could provide scope for selection. Polyploids are not common among the cultivated pulses, with the notable exceptions of the soybean and groundnut, indicating the strong genetical barriers between species. Nevertheless, Smartt and Haq (1972) attempted to produce amphidiploids between *P. vulgaris* and *P. coccineus*. They indicated that the production of amphidiploid beans would be extremely valuable potentially and might enable the culture of these grain legumes to be extended to environments in which production was not possible at the present. In support of their idea they cited the precedent of wheat in which polyploidy was followed by widespread dissemination and adaptation to widely different environments.

Until recently, one of the major obstacles to the utilization of wild species in pulse breeding was the lack of availability of germplasm in collections. In addition, relatively little emphasis has been placed on the grain legumes compared with cereal and root crops. This situation has changed fortunately and several of the international agricultural research centres have established germplasm collections. At CIAT, *Phaseolus* beans are being collected, evaluated and utilized in breeding. A regional network throughout Central America ensures that breeding lines are evaluated under conditions of the small-scale farmer, for whom the common bean is the major source of protein. At ICARDA, the emphasis is on lentils and broad beans (*Vicia faba*) for the dry areas of the Middle East and Africa, and at ICRISAT in India, germplasm and breeding activities are undertaken on pigeon peas (*Cajanus cajan*), chickpeas (*Cicer arietinum*) and groundnuts (*Arachis hypogaea*), all important pulses in the semi-arid tropics of the Indian sub-continent.

As Smartt *et al.* (1978) have pointed out for *Arachis* the exploitation of genetic diversity is complicated by genome differentiation, and further complicated by ploidy differences between the cultigen and diploid related species in the same section of the genus. In order to maximize the exploitation of the genetic resources of this section *Arachis*, it may be necessary to breed and select outside the confines of the cultivated species, firstly in order to concentrate the desirable factors to be introduced, and secondly to establish pathways by which transfer of genetic information to the cultivated species can be achieved. In other grain legumes where the exploitation of genetic diversity is equally difficult, such pre-breeding studies are an invaluable step in the process of utilizing genetic resources.

The international transfer of genetic resources and quarantine

The completely unrestricted transfer of plant genetic resources across international frontiers rarely happens. The transfer of plant materials on a global scale, either for utilization or for conservation of genetic resources, involves possible risks of widespread distribution of plant pathogens. Plant material may appear in the main to be free of pests or pathogens, but if not there may be accidental transfer. If these pests and pathogens are difficult to recognize they may affect the true expression of the genetic potential of the collected material. More importantly, non-indigenous pests and diseases are a serious threat to agriculture should they escape. Clearly these risks must be taken into consideration during the planning and execution of genetic resources activities, and may affect where materials will be sent for long-term storage. Most countries already have plant quarantine regulations which often constitute an important constraint to the movement of plant genetic resources.

The types of problems faced are to a certain extent, crop dependent, although several pests and diseases do attack more than one host. Vegetatively-propagated crops such as potatoes, yams and cassava present special problems for plant quarantine, because of the range of pests and diseases, especially systemically-borne diseases like viruses which can be carried in the vegetative propagules. This is not to discount the importance of seed-borne diseases especially viruses and viroids, but in general these are not transmitted with a high frequency during the sexual process.

Plant quarantine regulations may preclude the entry of certain species into a country, and therefore effect the total exclusion of the pest or disease, or they may allow release of plant materials after a suitable period of quarantine, during which tests are carried out for the detection of diseases. Quarantine precautions must be directed not only against species of pathogens, but equally important, against any specific pathogenic race. For example, among the seed-borne downy mildews there have been encountered some 40 pathogenic races of *Peronospora manshurica* and among the smut fungi, some 29 races of *Ustilago avenae* and 20 races of *U. tritici*. In addition to specific races related to vertical resistance, there are many races which show differences in aggresivity.

In centres of crop diversity, it is likely that such pathogens can be found, and this pathogen and pest variation must be taken into account when introducing seed from such regions.

The exchange of plant genetic resources generally does not involve large quantities of seed, and consequently samples cannot be taken for testing. It may be necessary to plant all the seed, and produce a new generation under quarantine. With respect to the genetic structure of the material, regeneration must be undertaken in such a way as to minimize the risk of genetic erosion. Perhaps this is not an aspect which is fully appreciated by plant quarantine officers.

The transfer of genetic resources in the form of tissue cultures from which pests and diseases have been eradicated, does provide the safest method of germplasm distribution. Nevertheless, tissue culture protocols are not available for all crops, and until these have been developed we are left with the problem of ensuring that pests and diseases are not disseminated around the world, and yet that genetic resources are fully available for utilization in plant breeding programmes.

7

Vegetables, Industrial Crops, Medicinal and Forage Plants

The greatest priority, as far as conservation is concerned needs to be assigned to the major food crops of the world, especially if it is determined that genetic erosion in any region has reached crisis point. But while the seven biggest tonnage crops (wheat, maize, rice, potatoes, barley, sweet potato and cassava) form the staple foods, and therefore rightly attract most attention, fresh vegetables and fruit are important in providing variety and supplying essential vitamins in most countries. There are other needs also, which demand that attention be given to species which are sources of industrial end-products, such as fibres, waxes, oils and gums, pharmaceuticals and even insecticides. Other categories of some importance are beverages and stimulants, and a very diverse range of species grouped together as forage plants. The view of genetic resources needs therefore to be a broad one, and should even include germplasm of wild species not previously considered for crop purposes: the search for species which are promising sources of new crops, should be considered.

Vegetables

Whether cooked or eaten raw, vegetables are prized throughout the world for their nutritional value and flavour. In many countries, particularly those still developing, they make up a substantial portion of the daily diet, while in other western nations nutritional evidence indicates that fibre intake should be dramatically increased, particularly through eating more vegetables. They provide not only vitamin and mineral nutrition, but in some cases considerable protein, in addition to dietary fibre.

In the tropics, a very large number of wild and cultivated plants is used as vegetables. There are about 2000 such species, but most of them are not well known. Many are not widely distributed and have limited potential. Nevertheless, taking tropical and temperate vegetables together, at least half a dozen of them are known to fall within the world's top 30 crop chart (see Table 2.3 p. 24) for total tonnage grown.

Vegetable crops have not, until recently, received the attention they deserve for a number of reasons. Invariably, total production of vegetables is underestimated, especially as vegetables are often grown in 'backyards' or even gathered within forests; these production figures rarely appear in official statistics. Often the value of vegetables as cash crops for small farmers has not been sufficiently realised, nor their nutritional value fully appreciated.

What of the availability of vegetable genetic resources? Within Europe at least, it can be appreciated that urban dwellers can seldom grow their own food, so that towns and cities need horticultural industries. Growers serving such needs have invariably selected vegetable types suited to particular local conditions. Over decades or centuries this has resulted in genetically discontinuous types of the same vegetables occurring within neighbouring areas. A very rich source of genetic diversity may well occur in small commercial companies supplying local vegetable seed (Crisp and Ford-Lloyd, 1981), as was discussed in Chapter 3. Such material needs to be conserved with some urgency for two reasons. Firstly, small companies are often ready for takeover by larger international companies, which are more interested in supplying varieties acceptable over a wide area, with the result that local types are superceded. Secondly, the practising breeder usually tries to use material as near as possible to his intended end-product, purely for ease of utilization. A diversity of genes within a crop-type is therefore of more use than the same or alternative genes in wild, or more primitive relatives. One of us has most successfully collected vegetable beets and Alliums in Italy, aided by the telephone directory 'yellow pages' to find the addresses of numerous small seed companies.

In the tropics, the problem of genetic erosion is fuelled particularly by the introduction of European-type vegetables, which are more prestigious than the local vegetables, slowly causing the latter to disappear. Throughout the world, there is also a desire by vegetable growers for F_1 hybrid varieties, which is further resulting in the replacement of landraces and open-pollinated cultivars. No matter to what extent any vegetable species is cultivated in any part of the world, there is evidently danger of rapid loss of important locally-adapted genotypes.

Tomatoes

The tomato is already the most important vegetable in many countries. Indeed, it leads the vegetables at number 16 in the top 30 crops chart. This is a very recent phenomenon and is largely due to the success of the plant breeder. The ancient herbalist Matthias de L'Obel back in 1581 had a different opinion. He stated that 'these apples are eaten by some Italians like melons, but the strong stinking smell gives one sufficient notice how unhealthful and evil they are to eat'. Evil and smelly, who knows? But unhealthy, definitely not! Its high level of consumption makes this crop one of the major sources of vitamins and minerals in many countries (FAO, 1980).

The origin and domestication of the tomato is not totally clear. The most likely ancestor of the cultivated tomato is the wild cherry tomato (*Lycopersicon esculentum* var. *cerasiforme*) which grows spontaneously in tropical and sub-tropical areas of the New World (Rick, 1976). It is more difficult to be specific about the time and place of domestication, which certainly took place before it was taken to Europe and Asia. Although the centre of diversity of the genus *Lycopersicon* is the Andean zone of South

America, there are reasons to believe that Mexico is the origin of domestication of tomatoes.

One thing is certain, and that is that the tomato is an excellent example of the great potential that local cultivars and wild relatives hold to improve resistance to environmental stress. Plant breeders over the past 30 years have made tremendous gains in yield, fruit setting, quality and ability to withstand pressures of handling and storage (Table 7.1). The ease with which tomato species can be crossed, together with a thorough knowledge of their cytogenetics, provide the breeders with tools with which to sculpture the genetic resources of *Lycopersicon* species, building them into cultivars of tomatoes. Still more genes have yet to be incorporated. Many genes favouring resistance have not been exploited. Wild relatives such as *L. cheesmanii* have been shown to possess tolerance to salty soils. *Lycopersicon pennellii* has the ability to withstand drought, while varied resistance to insect attack may be transferable from *L. hirsutum*, a relative with long glandular hairs and very strong odour. Fairly comprehensive collections of germplasm exist in China, Cuba, German Democratic Republic, Hungary, Netherlands, Peru, USA and USSR; they may hold valuable genes both recognized, and as yet undiscovered.

Table 7.1 Examples of germplasm utilization and genetic improvement in cultivated tomatoes. (From Esquinas-Alcazar, 1981.)

(a) Resistance to fungi
- *Alternaria solani* (early blight) from *L. esculentum* var. *cerasiforme*
- *Botrytis cinerea* (Botrytis mould) from *L. hirsutum*
- *Cladosporium fulvum* (leaf mould) from *L. pimpinellifolium*
- *Colletotrichum phomoides* (Anthracnose) from *L. esculentum* var. *cerasiforme*
- *Fusarium oxysporum* f. *lycopersici* (Fusarium wilt) from *L. pimpinellifolium*
- *Pyrenochaeta terrestris* (Corky root) from *L. peruvianum*
- *Septoria lycopersici* (Septoria leaf spot) from *L. hirsutum*
- *Stemphylium solani* (grey leaf spot) from *L. pimpinellifolium*
- *Verticillium albo-atrum* (Verticillium wilt) from *L. esculentum* var. *cerasiforme*

(b) Resistance to viruses
- Curly top virus from *L. chilense*
- Tobacco mosaic virus from *L. peruvianum*

(c) Resistance to nematodes
- Root-knot nematode (*Meloidogyne* sp.) from *L. peruvianum*

(d) Incorporation of the genes that prevent easy abscission of fruit and retention of undesirable pedicel stubs, from *L. cheesmanii*

(e) Increment of soluble solids from *L. chmielewskii*

Cucurbits

Cucumbers, squashes, gourds and melons, generally grouped together as cucurbits, are a vital feature of farm economies. They are extensively traded between nations but are also important in subsistence agriculture. In general, the fleshy cucurbit fruits have a poor nutritional composition, but there are some exceptions. The yellow-fleshed pumpkins and melons have a high ß-carotene content, the bitter gourd is high in iron and vitamin C and the winter squash contains much carbohydrate. In the tropics, the daily intake of these fruits tends to be high, so that their contribution of vitamins and minerals should not be underestimated.

Not only are cucurbits cultivated for their fleshy fruits, but also for their leaves, flowers, seeds and even roots. Only one species, *Telfairia occidentalis* is cultivated primarily as a leaf vegetable. Several species are cultivated for their dry seeds which are rich in protein and oil and are therefore comparable with groundnut and soya, for example the water melon, *Citrullus lanatus* and the bottle gourd, *Lagenaria siceraria*. The most important species in Africa for seed production is indeed *C. lanatus* from which oil is commercially extracted. The range of uses of different cucurbits is really quite amazing, for those species with fruits having a hard, tough rind are used for containers, cutlery, musical instruments and ornaments. This is not surprising as squashes, pumpkins and gourds have been associated with man in the Americas for at least 10 000 years. Recent evidence from radiocarbon dating of small rind remains establishes that squash had been introduced into eastern North America, probably from Mexico, before 7000 years BP, and well before any horticulture had been established involving indigenous plants (Conard *et al.*, 1984).

Citrullus lanatus is known to exist as at least 300 local and modern cultivars, many of which are particularly suitable for growing in the tropics. Although many local cultivars occur in S.E. Asia and India, the species originated in Africa, the truly wild forms growing mainly in the Kalahari Desert in South Africa. The plant is normally grown for its large yield of heavy, juicy, sweet fruits, although in Africa, local cultivars with bitter fruits are widely cultivated for the edible seeds. In West Africa, these seeds are called 'egusi', and are an interesting product of the traditional mixed cropping system, giving an additional crop with a high nutritional value. Some forms of water melon, known as citrons, have fruits with firm, hard, greenish flesh and large seeds. These are used for stock feeds.

Although resistance to various diseases has been obtained in modern cultivars by intensive hybridization between local and wild types (resistance to *Fusarium* wilt and different races of anthracnose), in general modern cultivars prove to be extremely susceptible to pests and diseases. There remains much to be achieved by looking to genetic resources to solve these disease problems.

Cucumis melo, the melon, muskmelon or cantaloupe is of importance more in developed than developing countries for its white-green juicy fruit. One form, however, (convar. *conomon*) is a pickling melon which is cultivated in China and S.E. Asia for the ripe fruits which are used as a cooked vegetable. The melon originated in tropical and sub-tropical Africa, where many wild forms occur. Breeding work has been initiated mainly in the USA where hybrid varieties with resistance to *Fusarium* wilt and to powdery mildew and downy mildew have been developed. There appears to be no significant risk that the wild and partly wild forms of *Cucumis melo* will disappear from tropical Africa; these may continue to form a gene pool from which disease resistance may be extracted.

Cucumis sativus, the cucumber, while having only minor calorific and nutritive value is second only to the tomato in importance as a vegetable in Western Europe. It is rarely cultivated in tropical Africa and America, but does hold some importance in Asia. De Candolle (1882) thought that it had been cultivated in India for over 7000 years, where tremendous variation has been found. This points to where the need lies for conservation, particularly as it is here that the putative ancestor, *Cucumis hardwickii* abounds. This is a wild form, small and bitter with sparse stiff spines, but which hybridizes easily with the cucumber, and is a possible source of useful genes.

The interesting point about cucumbers is that there is more genetic information available for cucumbers than for any other cucurbit (and indeed many other vegetables). More than 50 loci have been identified, and these include genes governing plant habit, sex expression, fertility, and several fruit characteristics. Clearly, here is a candidate for descriptors to be recorded at the gene level (see Chapter 5).

Cruciferous crops

Vegetables of the family Cruciferae are consumed in nearly every part of the world. This family includes cabbage, turnip, pak-choi, brussel sprout, mustards, cauliflower and kale. These may be harvested for their stems, buds, floral parts and seeds. In addition to their use as vegetables, they may also be grown for oilseed, forage and fodder, green manure and condiment (Table 7.2) Crucifers are the major vegetables of the diet of Chinese, Japanese, Koreans and Europeans; in China for instance, 50 per cent of the vegetables consumed are crucifers.

Brassica campestris includes the turnip which is important as a forage crop as well as a vegetable, and also several other subspecific forms representing leafy types developed for salad and pickling purposes in China and Japan. It is subspecies *rapifera* which is the true turnip, where the useful part is the storage organ, technically a hypocotyl. Subspecies *oleifera* or turnip-rape is the one most clearly related to the wild form (*eu-campestris*).

Next comes a wide range of subspecies mainly used for their leaves. Pak-choi or Chinese mustard (ssp. *chinensis*) is a leafy annual, of which the young blanched shoots are an important vegetable in China. Selection has produced some extreme forms with greatly enlarged leaf petioles and only narrow laminas. Pe-tsai or Chinese cabbage (ssp. *pekinensis*) forms a well developed head of leaves used in salads. Other leafy forms producing heads or rosettes of leaves are ssp. *narinosa* and ssp. *nipposinica*.

There are indications of two main centres of origin. The Mediterranean area is thought to be the primary centre for European forms, while eastern Afghanistan into Pakistan is considered to be another primary centre. More recently, in the early part of the nineteenth century in Britain, various forms of turnips and swedes (*B. napus*) were grown, often side by side. They crossed readily, and as a result it was impossible to obtain true-breeding strains. The breeding of hybrid varieties within *B. campestris* is an attractive possibility. It

Table 7.2 Cruciferous crops and their wild relatives. (From Toxopeus and Van Sloten, 1981.)

Taxonomic name	*Common name*	*Main use*
Brassica campestris		
B. campestris subsp. *chinensis*	Pak-choi	Vegetable
B. campestris subsp. *japonica*		Vegetable
B. campestris subsp. *narinosa*		Vegetable
B. campestris subsp. *oleifera*	Turnip rape	Oilseed
B. campestris subsp. *pekinensis*	Chinese cabbage	Vegetable
B. campestris subsp. *rapa*	Turnip	Vegetable/Fodder
Brassica carinata	Ethiopian mustard	Vegetable/Oilseed
Brassica juncea	Indian mustard	Oilseed
	Chinese mustard	Vegetable
Brassica napus		
B. napus var. *biennis*	Rapekale	Fodder
B. napus var. *napobrassica*	Swede turnip	Fodder/Food
B. napus var. *oleifera*	Oilseed rape	Oilseed
Brassica nigra	Black mustard	Condiment
Brassica oleracea	Wild relatives	
B. bourgaei (syn. *Sinapidendron bourgaei*) Canary Islands		
B. oleracea wild in England , Brittany (and northern Spain?); coastal		
B. robertiana northeastern Spain, mediterranean, France, (Italy?); coastal		
B. insularis Corsica, Sardinia, Tunisia; mostly coastal		
B. macrocarpa Egadi Islands near Sicily		
B. villosa-incana (complex) Sicily, mainland Italy, northwestern Yugoslavia; coastal		
B. cretica (complex) Greece and Aegean Islands, Crete		
B. hillarionis Cyprus		
B. oleracea var. *acephala*	Kale	Vegetable/Fodder
B. oleracea var. *alboglabra*	Chinese kale	Vegetable
B. oleracea var. *botrytis*	Cauliflower	Vegetable
B. oleracea var. *capitata*	Cabbage	Vegetable/Fodder
B. oleracea var. *gemmifera*	Brussels sprout	Vegetable
B. oleracea var. *gongyloides*	Kohl rabi	Vegetable/Fodder
B. oleracea var. *italica*	Broccoli	Vegetable
Raphanus sativus	Radish	Vegetable/Fodder
		Oilseed/Green manure
Raphanus spp.	Wild relatives	
Sinapis alba	White mustard	Oilseed/Green manure

is reported for instance that inter-subspecific hybrids of *oleifera* and *chinensis*, *dichotoma*, *narinosa* and *pekinensis* have given yields superior to the better parent. This gives an indication of the diversity of forms which need to be conserved for possible future utilization.

Brassica oleracea is a highly polymorphic species including many very distinct forms, each separated into botanical varieties by only one or a few gene differences. In these different forms almost all parts of the plant may be modified into storage organs which are utilized by man. In the various types of cabbage, numerous overlapping leaves surround the terminal bud; large axillary buds occur in the brussel sprout; kohlrabi is seen to have a swollen, bulb-like stem; and in the cauliflower and broccoli, the inflorescence and the flower buds are thickened and fleshy to give an edible 'curd'. Most of these forms are only of minor economic importance, being limited to temperate regions. The hearting cabbage (var. *capitata*) is however grown more extensively, and its importance is increasing rapidly in the tropics. In the Philippines, 10% of the budget spent on vegetables is used in the production of cabbage, which is a very popular vegetable.

The early evolution of the different cultivated types certainly took place in the Mediterranean area and possibly in Turkey. The leafy unbranched kales and branching thin-stemmed kales were the earliest cultivated brassicas, and gave rise to the forage kales. Later arrivals were both red and white cabbages (recorded in about 1150 AD). Herbals of the 16th century make reference to Savoy-type cabbages, as well as kohlrabi and forms of cauliflower. Sprouting broccoli was apparently first recorded in the 17th century along with the first 'sport' of brussel sprout.

The problems with regard to pests and diseases appear to be similar throughout all forms of brassicas. No one type can be specifically identified as a good source of resistance genes. Important resistance genes can be found in various cultivars and landraces scattered throughout the range of cultivated forms. For several species of brassicas loss of genetic diversity is considered critical, because of the replacement by hybrids of locally adapted cultivars. Some wild relatives have in the past been collected by plant breeders, but such efforts have not been systematic. Within the crucifer family as a whole are many isolated and genetically differentiated populations that are highly important as a source of useful genes.

Onions and their relatives

There are over 600 species described within the genus *Allium*. Only about a dozen of these are in use by man (see Jones and Mann, 1963 for a most comprehensive account) but represent a range of genetically distinct true species (Table 7.3). Of these, *Allium cepa* (onions and shallots) is probably the most important species in cultivation. It is an ancient crop and has been utilized in medicine and rituals as well as for food in Egypt since about 5200 BP and India since 2600 BP. It is only found in cultivation, although there are several wild species which come from areas of central Asia thought to represent the onion's centre of origin. Major seed collections of dry bulb onions exist in a number of countries throughout the world. However, the closely related and interfertile shallot (*A. cepa* var. *ascalonicum*) is much less well represented. There may be problems in the long-term conservation of cloves of shallots which do not set seed, but nevertheless attempts should be made to collect and maintain such germplasm as some of the characters in shallots could be valuable in onion breeding.

Allium fistulosum is the Japanese Bunching or Welsh onion which originated in China from an unknown progenitor. It is widely cultivated either for its leaf blades or for its blanched pseudo-stems (consisting of leaf sheaths). This species has been identified as an important source of resistance to a number of diseases affecting the dry bulb onion crop, and also possesses useful genes for winter hardiness and early flowering. These genes are considered of

Table 7.3 Cultivated *Allium* species and their characteristics. (From Astley, Innes and van der Meer, 1982.)

Species	*Common name*	*Bulb* formation*	*Basic chromosome number*	*Multiplication*
A. ampeloprasum	Great headed garlic	+	16 or 24	vegetatively
A. ampeloprasum	Kurrat	±	16	by seed
A. ampeloprasum	Leek	−	16	by seed
A. cepa	Shallot	+	8	vegetatively by seed
A. cepa	Onion	+	8	by seed vegetatively (northern USSR)
A. chinense	Rakkyo	+	16	vegetatively
A. fistulosum	Japanese bunching onion	−	8	by seed
A. sativum	Garlic	+	8	vegetatively
A. schoenoprasum	Chive	−	8 or 16	by seed vegetatively
A. tuberosum	Chinese chive	−	16	by seed

* + = yes; − = no; ± = occasionally

potential importance for onion breeding, and a number of plant breeders are now attempting hybridization and repeated backcrossing as a means of transferring the genes of importance. They are however substantially hindered by fairly high levels of sterility in the F_1s, even though the F_1 hybrids can be made with reasonable success, a fact well illustrated by the natural occurrence of species hybrids such as the so-called French and Egyptian 'shallot', and other vegetatively reproducing forms such as Beltsville Bunching Improved, *A. wakegi* and *A. proliferum*.

Garlic (*Allium sativum*) was originally grown in China and Europe as a condiment and for medicinal and ritualistic purposes. Its cloves are now used worldwide either fresh or dehydrated as a condiment, but in some regions, the leaves are used as well. Garlic is known only as a cultivated plant although it apparently has a closely related wild relative, *A. longicuspis* which may be the progenitor. The main problem of conservation of garlic lies with the fact that it only reproduces vegetatively. Nevertheless it is an extremely variable species, with different clones being adapted to growing under a wide variety of environmental conditions. Up until the present there has been little erosion of local garlic cultivars, but some attempts need to be made to identify and catalogue such cultivars. As Astley *et al.* (1982) point out, establishment of a world collection would be costly in time and finance and would be likely to have to involve *in vitro* conservation methods. In this case, maintenance of living collections in countries where the crop is important has something to recommend it.

The leek, kurrat and great-headed garlic are closely related forms belonging to the *A. ampeloprasum* complex. Historically, these cultivated forms can be traced to the civilisations of the Near East as far back as 4000 BP, but it is not possible to distinguish the groups to which these early forms belonged. Different regional cultural practices have enhanced the groups' divergence, and given us the different crops that we recognize today. Leek is an important crop in most of Europe; the edible part is the etiolated pseudo-stem formed by the leaf sheaths. Kurrat is similar to leek, but does not have as well-developed a pseudo-stem, so that its leaves are eaten fresh or cooked in areas of the Near and Middle East. It is very popular in Egypt where it is incorporated as chopped leaves with cooked beans in sandwiches. Great-headed garlic, also known as Elephant garlic is again a leek-like plant that produces large garlic-like cloves. It is grown in several countries throughout the world, including the USA, Greece, India and the Netherlands. Under certain growth conditions, only small bulblets are formed and these are named pearl onions which are used for pickling. Taking leeks alone, strong selection pressure for uniformity, tolerance to yellow stripe virus and winter hardiness may be narrowing the genetic base of European cultivars, but it should still be possible to collect and conserve more cultivars of this crop, as well as those of kurrat, great-headed garlic and wild forms of *A. ampeloprasum*. Improved knowledge of the variation within this species would provide valuable insight into the relationships between the cultivated groups, and could highlight important breeding material.

Looking at the genus as a whole, and at the utilization of wild species in particular, these present difficulties at every turn. Correct identification of the numerous species which have been described is often very difficult. Transfer of genes from these species to cultivated forms has in the majority of cases proved impossible because of the existence of strong species isolating mechanisms. Many characteristics of interest to plant breeders have been found in several wild species. For instance, *A. christophii* possesses strong resistance to neck rot, as does *A. galanthum*, while *A. commutatum* has proved resistant to leek rust. In the main, these await the development of techniques of genetic manipulation for their full exploitation.

Industrial crops

Direct consumption as food by humans is, of course, the most obvious and important use for plants. However, crops are not only grown to produce food directly, but also to provide more sophisticated end-products, particularly after the extraction and processing of the raw materials. Some plant products are of a complicated industrial nature requiring high technology to fabricate – paints, motor-car tyres and plastics, for example. Other derived products may be more closely related to the original plant, and to the food crops which have so far been under discussion. These include beverages, oil crops and sugar crops. They are often referred to as 'cash crops' because of their economic importance in international trade.

Beverages

Humans have always had a need to satisfy their thirst. Water remains the prime candidate for this, but since the beginnings of civilization, people have sought additives for water that might make it more tasty and perhaps even potent. Aside from fruit juices and alcoholic fermentations, coffee, tea, and to a lesser extent cocoa, are universally utilized as both flavourings and stimulants, and represent important cash crops for a number of countries.

Coffea arabica and *C. canephora* are the two species of commercial importance for the production of coffee. Arabica coffee is considered to have been domesticated first of all in the Yemen Arab Republic, but because it does not grow wild there, it is actually thought to have originated in Ethiopia. With the original spread of a limited range of germplasm first to the Yemen, and subsequently through the rest of the world, cultivars of arabica coffee now in use tend to have a very narrow genetic base. This can be partly illustrated by the devastation which has been caused recently by the two major diseases of coffee – coffee leaf rust (*Hemileia vastatrix*) and coffee berry disease (*Colleotrichum coffeanum*). Ethiopia, then, is the primary centre of genetic diversity for this species of coffee. It is the only country where arabica coffee grows wild as a forest species. It is also a prime example of genetic erosion (Frankel, 1973). The FAO Coffee Mission to Ethiopia in 1964 and 1965 explored a large part of Ethiopia looking for genetic diversity of this species. The mission collected a large number of samples from the main traditional source of seed (local forest coffee), but nevertheless found that their success was limited, as almost 90% of the forest cover of Ethiopia had vanished by the time of their visit. New roads, expanding agriculture and forest utilization were making in-roads into the remaining 10% leaving very little of the once unlimited germplasm sources. More recently the local coffee populations in Ethiopia have been under threat because of the introduction first of all of the coffee berry disease leading subsequently to the introduction of a few resistant genotypes from outside to replace the local susceptible landraces. All this adds up to disastrous genetic erosion and indicates an urgent need for effective measures to conserve any variation that is still left.

Canephora coffee has its primary centre of diversity in Central and Western Equatorial Africa and Madagascar. It seems that this species is very under-represented in genetic resource collections at the present time, even though it apparently shows increased tolerance to coffee leaf rust which may be transferred to offspring from interspecific crosses with *C. arabica* (Ferwerda, 1976).

Complications arise with conservation of the African species as it has been suggested that no systematic sampling method can be adopted under the conditions in which coffee grows in Africa. Conservation of both forms of coffee is further hampered by the inability to maintain viability of seed much longer than two years, making coffee germplasm conservation somewhat dependent upon the maintenance of collections of living plants. A major collection is maintained at the Centro Agronómico Tropical de Investigación y Enseñanza in Turrialba, Costa Rica.

Camellia sinensis (tea) is the most important beverage crop in the world. In India it occupies over 3.6 million ha and this constitutes over a third of the world's tea production. It is grown as a perennial crop for its leaves which are processed into the tea of commerce. It is known only in cultivation, for no true wild tea exists, and consequently its origins are in some doubt (Kingdon-Ward, 1950). Early plant exploration in north east India, Burma and southern China clearly demonstrated that so-called 'wild patches' of *assamica* tea reported in these regions are in fact abandoned patches left by migratory people in early times. These sources have directly or indirectly contributed planting material for over 60% of the world tea acreage. The introduction of the China variety (var. *chinensis*) has enriched the gene pool via the production of natural hybrids more suitable for commercial cultivation than either of the parents.

Tea germplasm collections in India ran concurrently with the introduction of seeds and plants from China at the beginning of the 19th century. Disorganized collection continued into the 20th century and in fact led to the discovery of the Assam variety of tea. Although tea populations in India have always been highly heterogeneous as a result of hybridization between the cultivated forms planted together in the early years of cultivation, the danger of losing valuable genes is now increasing. In India this is due to the uprooting of seed-grown sections of tea estates, clearing of jungles for new uses, and the popularity of a few superior vegetative clones resulting in the narrowing of the genetic base of populations (Singh, 1980). Collections of germplasm have been established in a number of countries, particularly India, Japan, Kenya and Malawi, and these may serve to alleviate the problem of genetic erosion in the future.

Theobroma cacao (cacao or cocoa) is one of the crops domesticated in Meso-America, probably in regions adjoining Guatemala and Mexico. Archaeological remains, possibly of Mayan origin, have been found in Guatemala, which show pods resembling 'cacao lagarto'. This variety may have been the first kind to be domesticated because of its soft husk and well-flavoured seeds. In Mexico, the same kind of cocoa is illustrated in ancient inscriptions, and its importance in religious ceremonies may be deduced from stone carvings of the Totonaca culture in Veracruz, which date from the 5th to the 9th centuries AD.

Cocoa is extremely variable with two subspecies being recognized, namely ssp. *cacao* and ssp. *sphaerocarpum*. The species can also be artificially classified from the point of view of the end-product, and this can give some guidance with regard to the utilization of its genetic resources. 'Criollo' cocoa is rated the highest in quality but is low-yielding and lacking in hardiness; very little is now available. Amazonian 'forastero' is much higher yielding and hardy, and today supplies the bulk of the world's cocoa. 'Trinitario' cocoa is essentially a hybrid form of criollo and forastero ancestry, and for that reason is much more variable in economically important characters. The trade refers to it as 'fine cocoa'.

Consumption of cocoa beans for the manufacture of drinking cocoa and chocolate has recently been high enough to exceed supply. It is therefore an important commercial crop, and worthy of conservation. Cocoa is highly genetically variable due to the existence of an outbreeding mechanism, and also to the fact that all subspecies and forms are interfertile. There are clearly many genotypes worth conserving. Worthy of report in this respect is a recent project aimed at collection and conservation of wild cocoa trees from the Amazon basin region of Ecuador. Called the 'London Cocoa Trade Amazon Project', it was initiated to collect material from this part of Ecuador, close to the centre of origin of the wild cocoa species, where it was already known that important genetic characteristics could be found. These include resistance to disease, especially witches broom, and advanced growth and yield characters. Over the last 10 or 15 years, oil field developments have led to roads being built into the region, and consequent colonization and forest clearance have put survival of the wild populations at risk. One positive effect of this though, has been the improved access to fairly impenetrable rainforest, making possible a more thorough and systematic programme of exploration and collection than had been possible before. Nevertheless, collecting cocoa germplasm is fraught with difficulties, some of which are outlined by Soria (1975) and Allen (1984). For instance, in the collecting areas, the labour of making the actual collections is slow and difficult, since the points of occurrence of the cocoa plants are unknown. Local Indians are helpful as guides, but apparently often confuse the sought-after forms with other species of *Theobroma*. Of course the greatest difficulty is presented by the short viability of cocoa seeds and budwood. Immediate transport has to be provided, to carry the material to its ultimate destination, otherwise it dies, making any expedition valueless.

Oil crops – oil palm

Plants which store oils or fats in their tissues are not in any way limited to specific taxonomic groups (Table 7.4). They belong to such distantly related families as Leguminosae, Euphorbiaceae, Compositae and Palmae, and cover both tropical and temperate regions of the world. They are consumed as food, and are put to a multitude of commercial and industrial uses. In terms of genetic conservation, some sources more than others have received recent urgent attention, while one species in particular (*Simmondsia chinensis* or 'jojoba') has lately been the subject of intense interest and exploitation partly as a result of world conservation strategies for threatened species of whales. Jojoba oil, for instance, is reported to be of high commercial value as a natural and superior replacement for sperm-whale oil.

It would be impossible here to report on the genetic status of all oil crops identified in Table 7.4. IBPGR has, of late, concerned itself with a number of these species, attempting to outline priorities for action. Nevertheless, it is worth taking one example in particular, which serves to illustrate a number of specific problems in terms of genetic resource conservation. The oil palm (*Elaeis guineensis*) is the most rapidly expanding plantation crop in the tropics. Malaysia has established itself as the main producer of oil palm, which was introduced to the Far East for the first time as an ornamental plant in 1848. Only four palms were planted in the botanical garden at Bogor, Indonesia, and these had originated from an unknown source in Africa (Arasu and Rajanaidu,

Table 7.4 Distribution of some main oil crops on a taxonomic and climatic basis. (© Crown Copyright, 1971. From Godin and Spensley, 1971.)

Species	*Family*	1	2	3	4	5	6	7	8
Soybean	Leguminosae	+				+		+	
Sunflower	Compositae					+	(+)	+	+
Groundnut	Leguminosae		+	+		+	+	+	
Cotton	Malvaceae			+	+I	+	+		
Rapeseed	Cruciferae			(+)		+		+	
Olive	Oleaceae				(+)		+		
Sesame	Pedaliaceae			+	(+)	+	+	+	
Maize	Gramineae	(+)		(+)	+	+	+I	+	
Safflower	Compositae			+			+		+
Coconut	Palmae	+	+	(+)					
Oil palm	Palmae	+							
Babassu palm	Palmae	+		+					
Linseed	Linaceae			(+)		+		+	
Castor oil plant	Euphorbiaceae			+		(+)		+	
Tung oil tree	Euphorbiaceae					+		+	
Oiticica	Rosaceae				+				
Niger seed	Compositae			+					
Hemp	Cannabaceae							+	
Perilla	Labiatae							+	
Poppy	Papaveraceae							+	+

( ) Limited cultivation possible; I, irrigation necessary. 1, tropical rain forest; 2, tropical monsoon; 3, tropical savanna; 4, dry tropical, i.e. steppe and desert; 5, humid subtropical; 6, dry subtropical, i.e. Mediterranean; 7, humid temperate; 8, dry temperate.

1975). Oil palm is certainly one of the most extreme examples of a crop with a narrow genetic base, for the seedlings derived from the four Bogor palms were subsequently planted at Deli in Sumatra. These gave rise to the Deli dura population which forms the backbone of the oil palm industry in the Far East. In fact, until the early 1960s, it formed the sole planting material. It is only recently that var. *tenera* material, originating as breeding stocks developed in Zaire, has been introduced into the breeding programmes of the Far East. However, Deli dura still continues to be used as a female source, and even on the male side, use of *tenera* material is still very limited.

With such a narrow genetic base, the extent and rate of genetic improvement through plant breeding is severely restricted. The introduction of new material into the gene pool is expected to enhance the scope of future breeding and selection. This particularly applies to oil palm which has a generation time of more than 10 years (Ooi and Rajanaidu, 1979). These are clear reasons for making sure that all existing genetic diversity be conserved for use in future breeding programmes. Even though genetic erosion threatens the centre of diversity which is in West Africa, it is necessary for Malaysia in particular to take the lead action, as it has more economic incentive to do so than African countries. Yet again, the major threat to the natural resources in Africa is the clearing of large tracts of land for development, but also a second threat comes from the practice of thinning palm groves to enable inter-cropping with food crops.

Sugar beet

Originally, all sugar came from sugar cane grown in the tropics, but nowadays beet sugar (from *Beta vulgaris*) accounts for nearly half the total world production of the refined product. Production of sugar beet is mainly in temperate countries, especially Europe, the USSR and North America.

Evidence concerning the first beets used by humans comes from linguistics (Ford-Lloyd and Williams, 1975) and this relates to vegetables not too far removed from the wild maritime forms of beets existing today. Forms used mainly for their leaves and certainly lacking any swollen root, may have resembled Swiss Chards, and these were described by both Aristotle (350 BC) and Theophrastus (300 BC). Much more recently, in 1554, Matthiolus gave one of the first descriptions of a cultivated beet having a swollen root. Soon after this Olivier de Serres referred to the sweet juice from red beets as 'semblable a syrop au sucre', and farmers even used fodder beets to produce syrup for feeding their bees (Bosemark, 1979). It was not until 1747 that the German chemist Marggraf studied the sweet substance in beet and found that it was in fact sucrose, which until then was thought to exist only in cane. Following this, Archard, a student of Marggraf made selections of beet types more suitable for extracting sucrose, and also built the first-ever sugar beet factory in 1802. Sugar beet can therefore be regarded as one of our youngest crops, with a history going back less than 200 years.

There has been some dispute as to exactly what beet material was originally used to select for the first true sugar beet, but three things are certain. First of all, during the course of only about 100 years leading up to 1858, the extractable sucrose percentage was raised from 1.56 to 13. Secondly, by 1858 the German plant breeder Knauer had developed the 'Beta Imperialis', usually considered to be the mother of modern sugar beet. Thirdly, despite the absence of barriers to hybridization between the cultivated and wild members of the section *Beta* and the availability of source populations with a wide range of genetic variation, the actual genetic base of sugar beet is probably narrower than that of most other cross-fertilizing crop species. In part at least, this has been a result of intense and one-sided selection for high sugar content. This is not to suggest that the present gene pool which includes sugar beet is particularly narrow, or in immediate danger of being exhausted. Traditional open-pollinated multigerm cultivars have been the subject of recent conservation activities, and undoubtedly represent considerable genetic variation. However, sugar beet breeders are even now attempting to broaden the genetic base even further, and certainly in the long run, according to Bosemark (1979), this may well be of importance in saving the crop from disease epidemics and maintaining progress in breeding.

How is this to be done? It is indeed lucky that if the broadening of the base is ever to include diverse forms and landraces of table beets and fodder beets, these forms have already been collected and conserved in their centre of diversity in Turkey. Williams and Ford-Lloyd (1975) reported that in 1973 these types of beets, having been grown and used locally for generations, were being replaced, even in backyard cultivation by modern European varieties of sugar beet.

Of greater difficulty in utilization, but under much less threat of extinction are diverse wild forms of beets. Nevertheless, as far back as 1932, Munerati persevered with *B. maritima* biotypes exhibiting *Cercospora* resistance, and by repeated backcrossing and selection, developed sugar beet populations showing the desired resistance. These have subsequently been used by other breeders, so that virtually all existing leaf-spot resistant cultivars in Europe and the United States have been derived in part from his material. Much more recently, more distantly related species such as *Beta patellaris* have received attention because of their strong resistance to beet cyst nematode. In the United States, Savitsky appears now to have come near to producing varietal material with this resistance (Savitsky, 1975), as have other scientists working in the Netherlands (Heijbroek *et al.*, 1983).

To highlight the fact that plant breeders' interests in genetic resources can never wane, a new syndrome of sugar beet has recently come into the news. This is called 'rhizomania'. It is a disease caused by beet necrotic yellow vein virus which is soil-borne, and is transmitted in the soil by the common fungus *Polymyxa betae*. It is reported that this disease has effects far more serious than those of any other disease affecting beets (Dunning, 1983). It is far too early at this stage to guess whether resistance in any form can be found in existing germplasm collections of beets. The fact that the effects of rhizomania were first felt in the Po River plains of Italy in the 1950s may provide a lead as to where to look for resistance genes; there may have been time for resistance to have evolved naturally. This is the region where Munerati found his *Cercospora* leaf-spot resistant wild beets. Could material of a similar origin provide the answer to rhizomania?

Medicinal plants

> '. . . and the fruit thereof shall be for meat, and the leaf thereof for medicine'.
>
> Ezekiel 47: 12.

Humans have used herbal medicines since the beginning of their civilization. Ancient Egyptians, Indians and Chinese used a variety of plant forms and products for curing all kinds of ailments. In ancient Britain it is likely that the Druids became acquainted with Greek medicine and the teaching of Hippocrates (the father of medicine) via the Phoenicians. Welsh legend has it that even before this time, perhaps as early as 2430 BP, botanical medicine was practiced in this part of the British Isles.

It is unrealistic, however, to use legends and ancient customs as evidence of the importance of plants in pharmacological and therapeutic technology, and even less so about any need for programmes of genetic resource conservation. It is much better to quote figures for recent use of therapeutic agents derived from plants. Over ten years ago, a survey in the United States listed a total of 76 different chemical compounds of known structure, derived from higher plants, and used at a high frequency in prescriptions (Farnsworth and Bingel, 1977). Only six of these were actually produced commercially by synthesis. Most commonly encountered higher plant extracts mentioned in this survey are belladona (*Atropa belladona*), ipecac (*Cephaelis ipecacuanha*), opium (*Papaver somniferum*), rauwolfia (*Rauvolfia serpentina*), cascara (*Rhamnus purshiana*) and digitalis (*Digitalis purpurea*). More recently in a review of anti-cancer agents of plant origin (Salunkhe *et al.*, 1984) over 20 classes of chemical compounds are listed which have some medical potential. These can be extracted from a range of plant families including the Celastraceae, Cruciferae, Nyssaceae, Ranunculaceae, Rutaceae and Taxaceae.

If one can be convinced of the importance of plant-derived products in medicine in the future, it is certainly worth turning back to look at ancient practices and customs for guidance. A large part of the population of India depends even at the present time on the ancient Indian systems of medicine, Ayurveda, Unani and Siddha. The greater part of the indigenous drugs used are from plants, and they may be represented by over 2300 different species. These are only now being screened and tested by modern techniques to accurately define their importance.

Leaving aside the considerable problems of confirming the importance of putative medicinal plants, what of the problems of genetic conservation? By and large, medicinal plants continue to be cultivated in the same ways as they have been grown for hundreds or thousands of years. As such, these crops are not intensively domesticated and have not acquired genes for high productivity in cultivation. The wild populations of these plants thus constitute the major genetic resource. Forests, natural reserves, uncultivated lands and backyards are the main habitats where genetic resources of medicinal plants are to be found, and the safety of these resources is linked with the existence of the habitats in which they occur. Types confined to 'backyard' cultivation may still be fairly safe. Those linked with forests are fast eroding, particularly in the tropics, and the destruction of ecosystems has accelerated genetic erosion.

Apart from the threat of genetic erosion, there are other problems related to the conservation of medicinal plant resources. Very little is known about their phytogeographical distribution patterns, modes of reproduction, breeding systems, storage potential of the propagules and so on, making their effective conservation more difficult. Also, major problems arise because of confusion regarding identification by common name of the different plant types. Even in adjacent localities in a particular country, native people may use a given plant for several ailments, and conversely a given ailment is often treated by several plants. Coupled with the likelihood that both the names of the plant and the ailment may change from region to region, this can cause much confusion at later stages of the conservation and utilization process.

Some medicinal plants of course are already highly commercially exploited, and can therefore be treated in a more straightforward manner when it comes to conservation. This applies particularly to the opium poppy, *Papaver somniferum*. The crude extract of opium latex yields morphine and codeine alkaloids which continue to be the most effective pain-relieving drugs known to mankind. The crop covers a large area of over 57 000 ha in India alone, and even in 1977 this resulted in 15 tons of opium latex being produced. Because of its undoubted commercial and medical importance, *Papaver somniferum* has been intensively collected during the early 1970s in Turkey. This was timed to beat the legislated ban on unofficial growing of opium poppy in Turkey, and resulted in a collection of over 2000 accessions now held in the genebank in Izmir. More recently smaller collections have been made in Rajasthan (northern India) and these have been characterized and evaluated for the presence of dwarfing genes and susceptibility to powdery and downy mildews, as well as general latex yield and morphine and codeine contents (Gupta *et al.*, 1980).

Likewise, species of *Rauvolfia* have received much attention. Their importance as medicinal plants lies in the discovery in 1952 of reserpine as a cure for high blood pressure. A subsequent steady decline of supplies of *Rauvolfia serpentina* roots from Asian countries has obliged western pharmaceutical companies to make a world-wide survey of *Rauvolfia* species. A richer source of the drug was found in roots of *R. vomitoria* from Central Africa. The same applies to the discovery of the effectiveness of cortisone for the treatment of rheumatoid arthritis in 1949 which led to a world-wide search for plants containing cortisone or similar compounds. This led to the further discovery of the cortico-steroid-containing *Dioscorea composita* from Mexico and *D. floribunda* from Central America. A similar hunt in India revealed *D. deltoidea* and *D. prazeri* as very rich sources of the drug diosgenin.

The specific examples cited above point to the need for collection and conservation of medicinal plants of proven worth. There is also a need to tackle the survey and conservation of lesser-known species of pharmacological potential. Medicinal plant resources have recently been described as 'a sleeping giant' (IUCN, 1980). It is to be hoped that he does not sleep for too long, else he might never wake up.

Forage grasses

Forage grasses present an interesting proposition from a number of points of view. On the one hand, grass must be one of the earliest plants used by man, but one of the most recent to be regarded as a crop. The history of domestication of most forage grasses goes back only over the last 80 years. The result of this is that many varieties still under production today are little differentiated from the ecological populations from which they arose. For similar reasons, even recent collections can occasionally uncover germplasm which is superior to that in current use, and which can be released as commercial cultivars with a minimum of further selection and breeding.

The reasons why the domestication of forage crops is generally so recent are also important to consider. They can be in part related to the general sufficiency of the naturally occurring pastures for grazing at low intensity. These pastures have also proved highly flexible due to their genetic heterogeneity, and have had the capacity to 'cope' with a move towards higher-intensity farming and higher fertilizer inputs in more recent times. It is of course this heterogeneity which it is important to conserve.

The other main reason why effective breeding of pasture grasses is comparatively new is that attempts at breeding 50 years ago were unsuccessful primarily because of a lack of knowledge of the overall genetic system of the plants concerned. The existence and effects of the outbreeding nature of these species was not appreciated, and likewise the 'conflict of interest' which arises between sexual and asexual reproduction. Strong opposing selection forces operate at these different reproductive phases. The early grass breeders have learned lessons of significance for the effective conservation and maintenance of forage genetic resources. Frankel and Hawkes (1975) have stressed the need for knowledge of population structure when determining conservation strategies, both in terms of sampling during collection and maintenance of germplasm. Nowhere is this more readily applicable than in forage genetic resources.

Staff at the Welsh Plant Breeding Station, Aberystwyth, UK, are foremost in determining the best strategies for conserving, maintaining (Fig. 7.1) and utilizing forage grass germplasm (Tyler, 1979; Tyler *et al.*, 1982). They have organized or participated in at least 17 expeditions over the last 20 years, covering many parts of Europe. They maintain a large seed collection under optimal conditions, and also distribute seed for utilization worldwide.

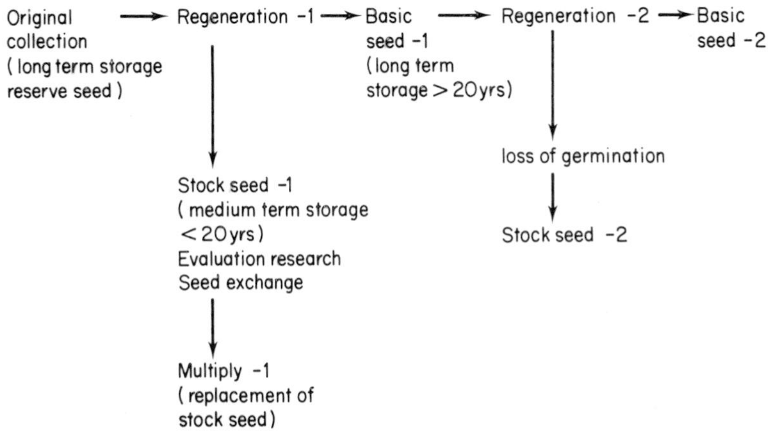

Fig. 7.1 Regeneration strategies for forage grasses. (From Breese and Tyler, 1981.)

Their procedures for the development of germplasm are precise and efficient. Germplasm samples are evaluated in field plots under both frequent and infrequent cutting regimes and in glasshouse and laboratory experiments. Elite populations are indentified which may show large differences in the expression of one or more desirable characters over control varieties, but show deficiencies in other characters, or appear to have an overall adaptation with one or more useful characteristics and no serious disadvantages compared with control varieties. These elite populations are then passed on to the breeders, either for direct utilization by selection or for incorporation into hybridization programmes.

As a result of these efforts, many forage grass varieties in current UK agriculture are derived from introduced gene sources, and these include

Table 7.5 Forage varieties in UK agriculture derived from introduced gene sources. (From Tyler *et al.*, 1982.)

Variety	*Origin*	*UK national list*	*Plant variety rights*
Perennial ryegrass			
Ba 9436	Zurich, Switzerland	Submission 1982	Submission 1982
Ba 9434	Pays de Herve, Belgium	Submission 1982	Submission 1982
Ba 9433	Ardennes, Belgium	Submission 1982	Submission 1982
Ba 9438	Pays de Herve, Belgium	Submission 1982	Submission 1982
Cocksfoot			
Saborto	Hybrid; indigenous *glomerata* 3 subsp. *lusitanica* ex Portugal	1972	1972
Calder	as Saborto with winter hardy genotypes of subsp. *lusitanica* ex Portugal	Removed 1978	Removed 1978
Cambria	ex Galicia N.W. Spain	1977	1977
Conrad	ex Galicia N.W. Spain	1980 (export only)	1980
Tall fescue			
Dovey	ex Durance Valley S.E. France	—	1978
Conway	ex Durance Valley S.E. France	—	1977
Italian ryegrass			
Trident	included Hungarian ecotypes	1977	1977
Bb 1906	Northern Plains, Italy	Submission 1982	Submission 1982

White clover			
Olwen	Durance Valley. S.E. France	1975	1977
Sabeda	Normandy N.W. France	1972	1977
Katrina	Indigenous var. 3 ecotypes ex Israel and Turkey	1976	1977
Donna	Scilly Isles	1977	1977
Ac 3053	Spain	Submission 1982	Submission 1982
Ac 3160	Italy	Submission 1982	Submission 1982

cultivars of perennial ryegrass (*Lolium perenne*), Italian ryegrass (*Lolium multiflorum*), tall fescue (*Festuca arundinacea*) and cocksfoot (*Dactylis glomerata*) (Table 7.5).

Changing agricultural practices are unquestionably eroding the naturally occurring resources of forage species in many parts of the world. Semi-natural grassland which is an historical result of intensive and expert management, is rapidly disappearing. Tyler *et al.* (1982) report that many of the sites in Europe which they visited some years ago, no longer exist in their original forms. Within the UK, consultations with the Nature Conservancy Council have resulted in the identification of indigenous ecotypes in danger of loss through changes in management and for other reasons. Water meadows and flood meadows in south and southwest England are a good example. Although a mass of naturally occurring forage germplasm must still be available to all concerned, the loss of ecotypes as outlined could result in a serious elimination of resources for future breeding activities. It is as well that some plant breeders continue to be effectively active in combating this threat.

8

Forest Genetic Resources

The scope and nature of the problem

The genetic base of most forest trees is generally broader than most of our food crops, but this is no cause for complacency. Undoubtedly forest genetic resources are under great threat. This comes not only from the ever-increasing timber industry world-wide, but in developing countries from increased population pressures leading to forest clearance for agricultural exploitation through shifting cultivation and for fuel. The scale of forest genetic resources is quite daunting, because of the large number of species of actual or potential economic importance, and the fact that these species are components of some of the largest and most complex ecosystems known to man. Forests comprise a wide range of vegetation formations from the sub-arctic to the tropics (Table 8.1). In each of these zones, they differ markedly in the way in which they have been exploited and also in their species richness and diversity, features which influence their usefulness to man. While it is clear that forests have been treated in the past as a non-renewable resource, this cannot continue. In the

Table 8.1 The distribution of forest area by major formation classes. (Adapted from Brazier *et al.*, 1976.)

Formation class	*Area (million ha)*	
Tropical zone	1456	
1. Wet evergreen		560
2. Moist deciduous		308
3. Dry deciduous		588
Subtropical zone	224	
4. Wet evergreen		8
5. Subtropical moist		20
6. Subtropical dry		196
Temperate zone	448	
7. Broadleaved forests 8. Coniferous forests		448
Boreal zone	672	
9. Boreal forests		672

temperate northern hemisphere for instance, deciduous hardwood forests have been drastically reduced in area ever since Neolithic man began to cut down the forests for agricultural exploitation of the land.

Forests and woodlands provide a rich variety of goods, useful to affluent industrial and poor rural communities alike: timber, sawnwood and panels for construction, walls, doors, shuttering and furniture; pulpwood for pulp, paper, cartons and rayon; poles, posts, mining timbers and railway track sleepers; fuelwood; fodder, fruits, game meat, honey, pharmaceuticals, fibres, resins, gums, dyes, skins, waxes and oils; beauty, amenity and recreation. Forests also influence local and regional climates, generally making them milder, and help to ensure a continuous flow of clean water. The careful management of watersheds is of increasing importance in many tropical countries, because the removal of the forest cover often leads to disastrous soil erosion. Tropical cloud forests increase the availability of water by intercepting moisture from the clouds, and forest removal can lead to drought. Regeneration of forests may be slow or even prevented if soil erosion has proceeded to any great extent. The widescale destruction of forest ecosystems has many other effects. If tropical forests are exploited, as generally they are, with scant regard for their ecological characteristics, the resource cannot renew itself. Temperate forests seem to have reached a state of equilibrium in terms of area, although they are still undergoing genetic impoverishment, but tropical forests are contracting rapidly, at the rate of some 110 000 km^2 $year^{-1}$. It is also apparent that the lowland forests are being destroyed at a faster rate than other tropical forests. This is particularly alarming, as scientists have yet to understand much about the nature of the tropical rain forest ecosystem, and all the different species, both plant and animal which are components of the system. The Amazon river basin is one of the most diverse floral reserves of the earth. It is, however, one of the least known taxonomically. This relative lack of knowledge, combined with the rapid pace of regional development, is especially threatening to the genetic variability of the economic and potentially economic species of forest, medicinal, vegetable, fruit and ornamental plants in the region.

In many developing countries there have been significant increases in recent years in the amount of forest clearance for the timber industry, and the subsequent planting of exotic species. While tropical forest ecosystems are inherently highly productive, the productivity of individual trees may be quite low. Foresters have therefore turned to exotic species which are considerably more productive than indigenous tree species.

A single method of forest genetic resources conservation cannot have universal application (Roche, 1975). Forest ecosystems are very diverse, and although theoretical considerations dictate which conservation options should be applied, practical constraints vary from country to country. In addition the biological and other information necessary for the implementation of adequate conservation strategies is often unavailable. In temperate zones, the methodology for the conservation of the genetic resources of many of the softwood and hardwood species is well developed, and a suitable infrastructure already exists to implement this methodology. The programmes in Canada for *Pinus* and *Picea* species, in California for a range of conifers such as species belonging to the genera *Pseudotsuga, Pinus, Sequoia*, and *Abies* amongst others, are well developed. But this is not generally the case in the tropics.

The importance of conserving and utilizing existing variation is recognized as fundamental in most tree species used in large-scale industrial plantations. However, little or no information is yet available on intra-specific variation in a large number of tropical trees which today are receiving increased attention as providers of goods and services for rural communities. Recent attention has also been given to the importance of genetic conservation of tree species which are used as a source of fuelwood in arid and semi-arid areas. In 1979, IBPGR agreed that it should concern itself with such species in these areas, as well as with tree species of value for planting by rural communities. By 1982, joint IBPGR/FAO collection and conservation in eight countries in the semi-arid areas had been initiated, primarily of *Acacia, Atriplex* and *Prosopis* species. Furthermore, seed of *Eucalyptus microtheca* has been distributed to IBPGR/FAO cooperators, so that self-sufficiency in fuelwood species in these areas of low rainfall could be achieved.

The importance of conservation of forest genetic resources becomes apparent when one realizes that the afforestation programmes of many tropical countries are dependent upon exotic species. For example, softwood forest depletion in Central America could have serious consequences elsewhere, since the pine species of this region provide the genetic raw materials for other regions.

The scope of the conservation problem is great because of the number of species involved, and the large areas needed for conservation. It is impossible to list all the important tree species, but some indication of the diversity of those which have a high conservation priority is given in Table 8.2. Some of these species, including the conifers are widely distributed in large populations which have become differentiated ecologically, so that intra-specific variation is a common feature. The diversity of the tropical rain forest is enormous. The breeding system affects the population structure, and this has a direct influence on the potential conservation strategy. For this reason, the gathering of this information is of importance in forest genetic resources conservation.

Breeding systems and population variation

The fact that forest trees have a long life cycle and that they are large are important considerations for their effective conservation and utilization. Richardson (1970) states that the size, stand structure and regenerative powers of forest tree species render them less liable to catastrophic elimination than most herbaceous species. Furthermore, the long life cycle ensures that by the time forest trees have reached maturity, they have experienced and survived a wide range of climatic conditions. Forests are geographically widespread, and it is clear that there has been local differentiation of tree populations, particularly of an ecotypic nature.

Temperate and tropical trees may have different variation patterns because of differences in their genetic systems and population dynamics. Most temperate trees that have been prominent in tree improvement programmes, for example conifers, are monoecious, anemophilous, widely outcrossed and have typically a high population density while tropical trees possess a wide variety of breeding systems, are almost exclusively zoophilous, and typically have a low population density. Specifically, tropical trees (as compared to temperate trees) are likely to show greater inter-population variation due to one or more of the following factors: (1) reduced gene flow between populations; (2) variation in selection intensity on a small geographical scale; (3) genetic drift due to low population density combined with limited gene dispersal.

An understanding of the fundamental basis of tree variability depends upon adequate taxonomic knowledge as well as a knowledge of their genetics and breeding systems. At present the amount of detailed information available for the species that comprise the tropical rain forests, for instance, is still rather scarce. Nevertheless, research has commenced in several parts of the world, and Ashton (1976) has outlined a research strategy to study the Malayan rain forests. The objectives of this research include observations on the reproductive biology and population genetics of natural rain forest tree populations. The Malayan lowland mixed dipterocarp forest is probably the most floristically diverse of all terrestrial ecosystems (Poore, 1968). Whitmore (1972) has estimated that there are some 4100 tree species in 760 genera and 94 families in all types of vegetation in Malaya, and this gives some indication of the scale of the problem facing the genetic conservationist.

Many of the species of the West Malesian forests are dioecious trees (Ashton, 1976), and this has also been confirmed for the Central American semi-deciduous forests, a phenomenon which favours the maintenance of cross-pollination. Pollination problems due to low species densities, and synchronous flowering both affect the population structure and its variability. It is clear that among the dipterocarps one finds clinal variation, in species-poor forests, or discontinuous, allopatric variation over large areas, and this variation is often ecotypic (Ashton, 1976). Studies on the pollination biology of rain forest species include the anatomy of inflorescence initiation, phenology of inflorescence development, flower and fruit development and fall, the differential rates of abortion and mortality from bud-initiation to seedling establishment and their causes and the identity of insect and other visitors. In order to estimate the degree of genetical variation in tree populations, the technique of isoenzyme analysis, by starch gel electrophoresis, or polyacrylamide gel electrophoresis, has been applied in many genera. In this way variation in isoenzymes can be correlated with morphological variation, and the genotypic similarity of trees can be compared with their spatial distance.

It is impossible in a chapter of this nature to cover all the possible tree species of economic importance. For brevity, we have chosen only a few species as examples.

The tropical pines are fast growing conifers which produce large volumes of material suitable for timber, pulp and chemical products (Burley, 1976), and they have become planted as exotics on a widespread basis throughout the tropics and subtropics. The most important or promising commercial pines are six from Central America and the Caribbean, namely *Pinus caribaea*, *P. cubensis*, *P. occidentalis*, *P. oocarpa*, *P. patula* and *P. pseudostrobus*, and two species from S.E. Asia, *P. kesiya* and *P. merkusii*.

Table 8.2 Important forestry species with a high conservation priority. (Adapted from FAO, 1981.)

Region	*Species*
Mexico	*Cedrela odorata* (E)
	Picea chihuahuana
	Picea mexicana
	Pinus maximartinezii
	Pinus oocarpa
	Pinus patula
	Pinus radiata (E)
	Pinus strobus var. *chiapensis*
	Pinus tenuifolia
	Populus spp.
	Pseudotsuga spp. (E)
	Swietenia macrophylla
Caribbean, Central and South America	*Araucaria angustifolia* (E)
	Bombacopsis quinatum (E)
	Cedrela odorata
	Pinus caribaea
	Pinus oocarpa
	Pinus pseudostrobus (E)
	Pinus strobus var. *chiapensis* (E)
	Swietenia humilis
	Swietenia macrophylla
	Swietenia mahagoni
Mediterranean region, Southern Europe and Near East	*Abies nebrodensis* (E)
	Cedrus libani (E)
	Cupressus dupreziana (E)
	Pinus brutia (E)
	Pinus eldarica (E)
	Pinus pinaster
	Ulmus wallichiana (E)
North, North-East and Central Asia	*Pinus armandii* (E)
	Pinus koraiensis
	Pinus pentaphylla (E)
South and East Asia	*Araucaria cunninghamii* (E)
	Dalbergia latifolia
	Dipterocarpus spp.
	Eucalyptus deglupta
	Gmelina arborea
	Pinus merkusii
	Tectona grandis
	Tectona hamiltoniana (E)
	Tectona philippinensis (E)
Africa	*Acacia* spp.
	Aucoumea klaineana (E)
	Chlorophora excelsa (E)
	Diospyros spp. (E)
	Entandrophragma spp.
	Khaya senegalensis (E)
	Pericopsis elata (E)
	Terminalia superba
Australia	*Eucalyptus* spp.

E = endangered

Provenances, i.e. original geographic sources of seed, of *P. caribaea* and *P. kesiya* most commonly used come from Belize and the Philippines respectively, and are adapted to and productive in a wide range of environmental and management conditions. Variation in adaptive characters, such as germination rate, time to produce secondary needles, and survival at various stages does occur between populations of the different species, but it has not been possible to correlate this pattern consistently with environmental variation, when provenances have been grown in uniform trials.

An understanding of the breeding system of tropical pines, as with all trees, yields information on the amount and availability of seed at various times of the year. The nature of population structure and gene distribution will determine the conservation strategy to be implemented. Tropical pines are monoecious and wind-pollinated. The amount of pollen transport over distance may be a crucial factor in inter-population differentiation. Such information may be difficult to obtain, but nevertheless, according to Burley (1976), it is essential to know the effective population size and the amount of inbreeding to plan provenance tests, conservation measures and breeding programmes.

While the special importance of the tropical pines has been highlighted, one must not forget other important softwood species such as *Araucaria angustifolia*, which has been seriously depleted in southern Brazil, the tropical hardwoods such as teak, *Tectona grandis* and *Gmelina arborea* (Verbenaceae) and mahogany, *Swietenia mahogoni*, and the eucalypts. The importance of a species such as *T. grandis* needs no emphasizing, because of the value of the timber. This species shows considerable ecotypic variation, but there appears to be a centre of diversity in the Indian sub-continent, although the best teak timber is produced in Burma and Thailand. From a genetic conservation point of view, the trees from the Indian sub-continent are an important resource. The importance of the genus *Eucalyptus* is widely recognized. Half of the man-made forests in Latin America are of eucalypts, and approximately one third of those in Africa. The forests of northern Australia, New Guinea, Indonesia and the Philippines are the sources of the tropical eucalypts, and these show significant geographic variation.

Conservation strategies

As with other crops, there is the choice of both *in situ* and *ex situ* conservation. Although detailed information such as that outlined earlier in this chapter is needed to determine the number and sizes of tree populations to be conserved, conservation measures must often be undertaken before this is known. In his consideration of tropical pines, Burley (1976) has stated that for species of low, immediate, economic importance, conservation may be restricted to ecosystem conservation *in situ* and haphazard introductions *ex situ*. For the more important species, urgent measures should be taken to conserve both *in situ* and *ex situ* as many representatives of as many populations as practically and politically possible. In this way both static and dynamic conservation of genotype and gene frequencies will be prolonged. Burley's comment could well be applied to other groups of species as well.

In situ *conservation*

In terms of the methodology of conservation, forest trees have a number of distinguishing characteristics, among which is the fact that the majority are still in the wild state. Conservation of forest genetic resources must be considered as a dynamic component of plans for the management and utilization of a renewable natural resource, and must be compatible with other conservation objectives, with respect to wildlife, watersheds and protection of land against erosion. Nature conservation can result in gene pool conservation of constituent species. Its efficacy for this purpose is closely related to the size, number and distribution of nature reserves being protected. Scientific opinion suggests that representative samples of most forest ecosystems can be maintained if given effective protection in areas within the range of 100 to 1000 ha. If the species is widespread latitudinally and altitudinally, as are many temperate conifers such as *Pinus contorta* in North America, and it also shows considerable ecological variation, then several reserves must be established to conserve the genetic resources *in situ*. Frankel (1970) has indicated that *in situ* conservation is the ideal model for long-term conservation of forest trees. They are then considered as communities in balance with a stable environment, the stability being subject to the general vagaries of natural environments.

Again, let us turn to the three examples given earlier to illustrate some of the details of conservation strategies.

Tropical pines

Kemp (1975) has outlined the importance of the Central American pines, and that the systematic investigation of variation through provenance research will take many years to complete. The destruction of these natural pine forests has intensified in recent decades, and some are threatened with extinction. In world terms, the pines are of utmost importance, but their conservation receives relatively low priority at a national level in Central America.

Individual genotypes will undoubtedly disappear as the pine forests are threatened by several factors, including exploitation for timber, land clearance for agriculture, fire, pests such as the bark beetle (*Dendroctonus* spp.), and seed and resin collection. However, it is unlikely that gene pools will be completely eliminated. The wide ecological and geographical range of the species has resulted in differentiation, and populations at the periphery of the distribution of the species are in need of greater protection. Managerial control of the forests is necessary to ensure their regeneration, through fire prevention measures, and this has been achieved with some success in the Mountain Pine Ridge of Belize and in Nicaragua.

In situ conservation of these tropical pines presents problems other than technical, namely social, political and economic. Where population pressures are increasing, the possibilities of *in situ* conservation are more limited. Forest management is a long-term consideration, and schemes, once implemented, must be maintained to allow the forests to regenerate. *In situ* conservation must also take into consideration the latitudinal and altitudinal range of the species. The actual size needed for individual conservation areas has been calculated by Kemp (1975) as 100 ha, with 40 ha containing some 8000–12 000 breeding individuals, while allowing for 60 ha of immature or regenerating stands.

Tropical hardwoods

The number of species of tropical hardwoods is vast, and their ecosystems occur throughout the tropics. Conservation *in situ* by protection of the ecosystem is particularly valid when trees are components of climax forest ecosystems and are species which do not regenerate adequately after logging. Because of the low densities at which these species often occur in tropical forest ecosystems, foresters have little information about what constitutes an effective breeding population, and it is extremely difficult to determine the optimum size of area for *in situ* conservation. Consequently, all that can be done is to ensure that the endangered species is represented by as many stems as possible in the forest ecosystems scheduled for allocation to strict natural reserves, or national parks. Estimates from as low as 200 to over 10 000 individuals have been suggested for *in situ* conservation. Kemp *et al.* (1976) suggest that even though a low number of individuals may be adequate, because of their capability of self-fertilization, the area of natural forest required may still be large because of the low and uneven stocking level.

Eucalypts

The conservation of *Eucalyptus* gene resources may be approached in two ways, either by the preservation of representative natural stands, or the establishment of populations. Natural regeneration of eucalypt forests in Australia is a common feature, and consequently in forest areas the gene pool is largely intact (Larsen and Cromer, 1970). Nevertheless, it has been recommended that seed trees be brought together in collections and propagated vegetatively in special clone orchards. From a genetic conservation point of view, it is important to collect not only from trees of proven provenance, which combine the most favourable characteristics, but also from populations as a whole so that the integrity of the gene pool is maintained.

Ex situ *conservation in the tropics*

The lack of silvicultural information on many of the tropical hardwoods precludes adequate *ex situ* conservation measures. Endangered populations should be propagated vegetatively in clone banks and from seed in order to ensure their survival. The establishment of plantations both within and outside the country of origin is possible for a few species, such as *Tectona grandis* and *Gmelina arborea*. One of the problems facing the conservation of tropical forest trees is that often their seeds are recalcitrant. However, seeds of some species can be stored at a low temperature, and pollen storage may even be a possibility.

Ex situ conservation may be relevant in cases where *in situ* conservation of ecosystems proves impracticable. Guldager (1975) has considered four alternative objectives for *ex situ* conservation:

(*i*) to establish and maintain conservation stands characterized as far as possible by the same genotype frequencies as the original populations (provenance) – static conservation of genotypes;

(*ii*) to establish and maintain conservation stands characterized as far as possible by the same gene frequencies as the original populations, thus avoiding total loss of any allele – static conservation of gene pools;

(*iii*) to establish conservation stands where gene frequencies are allowed to change freely according to natural selective forces – evolutionary conservation;

(*iv*) to establish conservation stands in which gene frequencies are deliberately changed by man in order to conserve characteristics important for plantation economy in a region and at the same time eliminate undesirable characteristics – selective conservation.

The maintenance of genetic integrity in *ex situ* conservation will depend upon the sampling of genotypes in the original population, the survival and growth of these genotypes, and the mating which occurs between them. Each of these is affected by a number of factors, many of which are common to the collection and maintenance of all genetic resources. Sampling errors apply equally to trees as they do to other crops, but the size of the population (or sub-population) to be sampled may be much more difficult to determine in forest trees on account of their size. The survival of genotypes in *ex situ* stands can be increased by careful planting and management so that competition between individuals is reduced. The long-term conservation of genes and gene frequencies is the most difficult part of the conservation programme. Conservation can be ensured for 30–50 years, but beyond this, as trees begin to flower then the accurate passing on of genetic information from a first generation *ex situ* conservation stand to the next depends on the type of mating, genetic drift and gene migration. Little can be done to ensure random mating, because of the seeding characteristics of individual trees, but migration in terms of pollen contamination can be minimized by adequate isolation. For example, studies carried out by Geary (1970) in Zambia on *Pinus kesiya* and *P. patula* showed that significant amounts of pollen could travel up to 2 km, so that seed orchards or *ex situ* conservation stands would need this degree of isolation. The problems of genetic drift and inbreeding can be minimized by planting 'large' stands, and in these, it is assumed that both factors will be of little importance for 2–3 generations, which may be as long as 60–90 years. This highlights one of the major problems facing the conservation of forest genetic resources, as living collections. It is not possible to see one's mistakes, as it were, for several decades, and then perhaps it may be too late. The importance of stability of management of such *ex situ* conservation stands, and the accompanying data management, cannot be stressed too highly.

Where to establish *ex situ* conservation stands will be determined by the selection of sites which provide optimum conditions for seedling growth and survival. But many of the tropical plantations are still too young to be assessed in terms of seed production.

Undoubtedly, where possible, seed and pollen storage provides the best method of conservation of forest genetic resources. But as outlined earlier and in Chapter 4, such technology is not appropriate for many species, such as the majority of conifers and hardwoods, in which seed is only short-lived in storage. The life-span of tree pollen is shorter than that of most tree seed under current drying techniques and storage conditions.

The future

It is obvious that *in situ* conservation offers the best strategy for forest genetic resources. In terms of gene pool conservation, Richardson (1970) has indicated that in our efforts we must encompass not only natural forests, but also artificial plantations of both indigenous and exotic species, seed stands, arboreta, provenance collections and clone archives.

Undoubtedly measures must be taken to reduce the destruction of natural forest for timber. The planting of fast-growing species in the tropics, such as *Albizia* (Leguminosae) which provides valuable timber, could be encouraged rather than further cutting down of the natural forest. Agricultural exploitation of the land should not be continued to the extent that environmental disturbance becomes irreversible. Furthermore, the exploitation of trees as sources of fuelwood needs greater research input, to identify species which are tolerant of drought conditions, but which are fast-growing. Environmental pollution leading to acid rain is of great concern in many industrial countries of the developed world, causing destruction of areas of conifer forest in central and northern Europe. *In situ* conservation stands of trees are threatened in this way, and if the problem persists, it will be necessary to increase, where possible, the conservation of genetic resources in seed banks.

9

Plant Genetic Resources – Future Prospects

While it is clear that IBPGR has played a pivotal role in recent years in the coordination of germplasm activities throughout the world, the success of such programmes will be achieved only through international collaboration. This international aspect to germplasm activities is both one of its strengths and weaknesses. Without international collaboration it will not be possible to undertake many important germplasm collecting missions, and certainly such activities generally come well down the list of many developing country priorities. Consequently the need for international finance is apparent, which comes mainly from the so-called developed, industrialized nations of the 'North'.

The Food and Agriculture Organization of the United Nations established its Unit of Crop Ecology and Genetic Resources in 1968, although interest was first raised in 1947. By bringing together panels of experts, it was possible to identify areas of the world where the threat of genetic erosion was greatest. However there was little or no financial commitment by FAO to the cause of genetic resources, and it was not until IBPGR was formed in 1974 that all responsibility was devolved to it and genetic resources activities placed on a sounder financial footing. As Hawkes (1983) has indicated, this period prior to the foundation of IBPGR does not signify time lost, because it had been possible to devise conservation plans, including details of seed storage and its costs during that period.

IBPGR receives its funding through the Consultative Group on International Agricultural Research, which itself was founded in 1971 under the joint sponsorship of the World Bank, FAO and the United Nations Development Programme (UNDP). The actual funding comes from several sources such as the Rockefeller, Ford and Kellogg Foundations, Regional Development Banks and donor countries. Its remit is to promote an international network of genetic resources centres to further the collection, conservation, documentation, evaluation and use of plant germplasm and thereby contribute to raising the standard of living and welfare of people throughout the world. It has helped to promote regional collaboration in genetic resources activities, through IBPGR regional coordinators. Funding has been provided to collect germplasm in various parts of the world, and it has sponsored over 260 collecting missions, especially in the high priority regions of the Mediterranean and the Near East, and in the drier parts of West Africa.

Considerable support has been given to regional activities in South East Asia, and regional coordinators are now in place in most parts of the world.

Political considerations

While the policies of IBPGR during the last decade may not be totally devoid of criticism, its work really has achieved a great deal in preserving rapidly disappearing genetic stocks on a worldwide scale. Moreover it has moved towards a phase of activity aimed at promoting the free exchange of conserved germplasm and to support its utilizaton in plant breeding programmes throughout the world. In addition, it continues to support training activities aimed at ensuring that trained personnel are available to undertake the task of germplasm conservation at national level. This educational development is essential for the successful implementation of germplasm activities worldwide. Unfortunately, genetic erosion continues, and conservation cannot be delayed until this educational development is completed.

In many parts of the world, facilities for the secure conservation of seeds and other propagules do not exist. It has been necessary to make arrangements for conservation. IBPGR has collaborated with several genebanks in the industrialized nations, and this has led to considerable political controversy in FAO concerning the location of major genebanks in the 'North' as opposed to the 'South'. While it is clear that the USA and USSR hold major germplasm collections, it should be stressed that these collections were established several decades ago, before the recent interest in genetic conservation. In fact, the collection held at the All-Union Institute of Plant Industry in Leningrad was established when Vavilov and his colleagues first undertook their original collecting missions. IBPGR has sought to redress the balance through the formation of a network of national and international genebanks.

Another political 'hot potato' relates to the idea that genetic resources should be repatriated to the countries where they were collected. Recently the industrialized nations have been accused by Mooney (1983) of what might be best described as 'germplasm hegemony' because of the large collections of plant genetic resources which they hold. They have been criticized for exploiting the gene-rich countries of the 'South'. We believe that these countries should unquestionably have free access to their own resources. It is also clear that the major crop improvement programmes are based in the developed nations. Such plant breeding programmes need continual sources of genetic diversity which must ultimately come from plant genetic resources. It is not realistic to expect plant breeding companies to make their commercial varieties available free of charge to the developing nations. Plant breeding is a capital intensive industry, and the production of a new variety may take up to 10 years or more. These companies are accused of accelerating genetic erosion. While it is true that the cultivation of modern varieties exacerbates the problem of genetic erosion, most of these varieties of the major staple crops, with perhaps the exception of the potato, bred in North America and Europe, do not find their way back to developing countries. Germplasm is utilized for solving problems in North America and Europe. Improvements in wheat, maize and rice, as well as potatoes, and other staple crops, are being made by the international agricultural research centres which have a mandate for research on these crops (i.e. CIMMYT, IRRI and CIP), based on the large germplasm collections which they hold. Such improvements cannot be made without collaboration with the industrialized countries, although the mandate of the IARCs is for agricultural development to benefit the less developed nations.

When IBPGR has sponsored collecting missions, a portion of the material collected has been deposited in a designated genebank for long-term safe keeping. If genetic resources are to be repatriated with any effective purpose, then it is essential that adequate preservation facilities and trained personnel are available to undertake the tasks at hand. Unfortunately in many of the countries where genetic diversity exists adequate storage facilities do not exist.

Recently there have been moves in FAO to provide a legalized framework for genetic resources activities, because several developing nations have felt that their interests are being compromised. Among their concerns has been the location of the genebanks in the developed nations, and the repatriation of genetic resources. We suggest that it is better to have collected genetic resources before they finally disappear, and have them safely deposited in genebanks, albeit in the 'North', than to have procrastinated while political considerations are resolved at length. Through this political debate, however, the problems related to plant genetic conservation are becoming appreciated by the public at large. No longer does 'conservation' just mean the preservation of a few rare species (which are probably well on the way to extinction anyway) or certain ecosystems, such as the tropical rainforest, but there is growing concern that the future of mankind is linked with a continuing source of genetic diversity. This diversity cannot be created overnight; the combinations of genes which have accumulated over thousands of years can be lost forever unless we take measures to prevent this. Surely our children will judge us for our success (or lack of it) over the future of plant genetic resources?

New procedures for utilization of genetic resources

It is a certainty that if plant genetic resources are not, and have not been gathered and preserved, then their continued existence is unlikely. If genetic erosion is not counteracted by adequate measures of conservation, then the future of plant genetic resources cannot be guaranteed. The effectiveness of the plant breeding process and consequently the condition of the world human population becomes a matter of much concern. Assuming that measures for adequately conserving plant genes continue on a broad scale and that useful genes or gene combinations can be identified and located, what new techniques are likely to be available for their full exploitation in the future?

The broad term 'genetic engineering' and the even wider term 'biotechnology' are now used by some to cover most novel approaches to plant breeding, particularly if they involve molecular biological techniques for gene manipulation and DNA recombination, or if they include *in vitro* techniques of cell and tissue culture. Some will regard these as being virtually synonymous with plant breeding (Valentine, 1978). Undoubtedly the ultimate aim is the same – to produce better adapted, more useful varieties of crop plants.

Whereas such techniques as pollen and anther culture are procedures which will assist the plant breeder in a fairly conventional way by speeding-up his work of identifying the first products of segregation and recombination from heterozygous F_1s, other new methods will allow new genetic combinations to be produced. These will inevitably affect the normal processes of plant breeding. More importantly, they will enhance the value and potential of plant genetic resources.

Amongst the most exciting of these is the aim to insert new genes into the chromosomes of plants such that the new genes are stably inherited and modify the properties of the plants. In programmes of genetic transformation involving the insertion of a purified gene, one special feature is that genes from virtually any organism can be used. The plant breeders are therefore not limited to genes already present in the crop under development but can search through many sources of genes to utilize. Once these programmes, at the moment very much in their infancy, are developed and perfected, then the onus will be on the people who look after plant genetic resources to evaluate material and identify genes ready for insertion. Progress to be reported so far comes from the Plant Breeding Institute in Cambridge, UK, where as a model system, a bacterial gene conferring antibiotic resistance has been 'engineered' and inserted into tobacco, making the plant resistant to antibiotic (Bevan *et al.*, 1983).

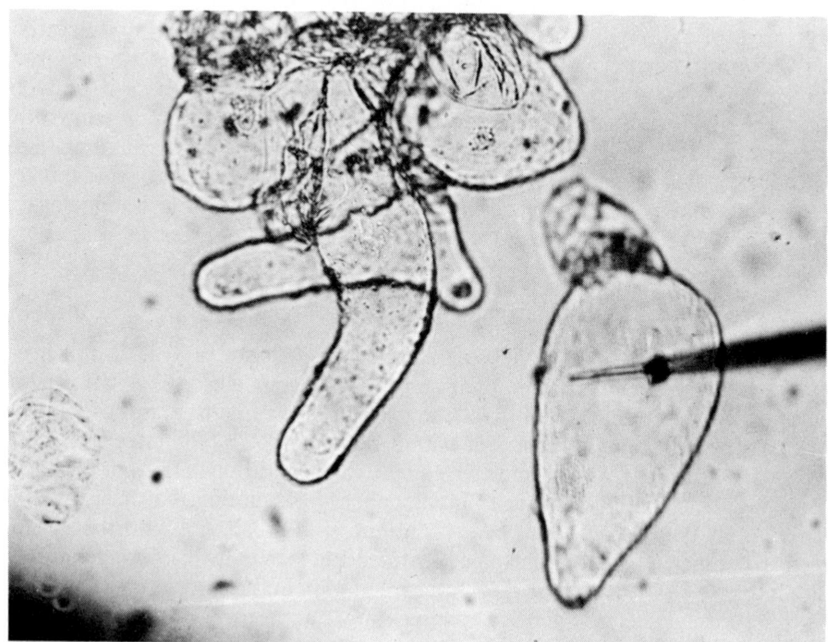

Fig. 9.1 Micro-injection of a single callus cell of sugar beet in culture.

Many approaches of this nature currently employ *Agrobacterium* as a vector for introducing new nuclear genes into dicotyledonous plants. This has proved much more difficult to use with monocots so that people are now investigating the possibility of directly micro-injecting various plant tissues or totipotent cells with isolated DNA or nuclear genes (Flavell and Mathias, 1984), or even non-nuclear genes contained in organelles (Fig. 9.1)

Another fascinating prospect for combining genes from widely separated sources other than by normal sexual means is protoplast fusion (Cocking and Riley, 1981). Here, the cell walls of various types of plant cells can be removed enzymatically. In a growing number of cases, these 'naked' protoplasts can be cultured, persuaded to form new cell walls, and to divide to form a callus from which new plants can be derived (Fig. 9.2). It is also possible that protoplasts from two genetically different sources can be made to fuse. This fusion will certainly result in a cytoplasmic hybrid or 'cybrid' and it is also possible that some degree of genetic recombination may occur between the two separate nuclei. So in the long run a new range of hybrid material may become available to the plant breeder and useful alien genetic variation, from hitherto inaccessible sources, introduced into crop plants.

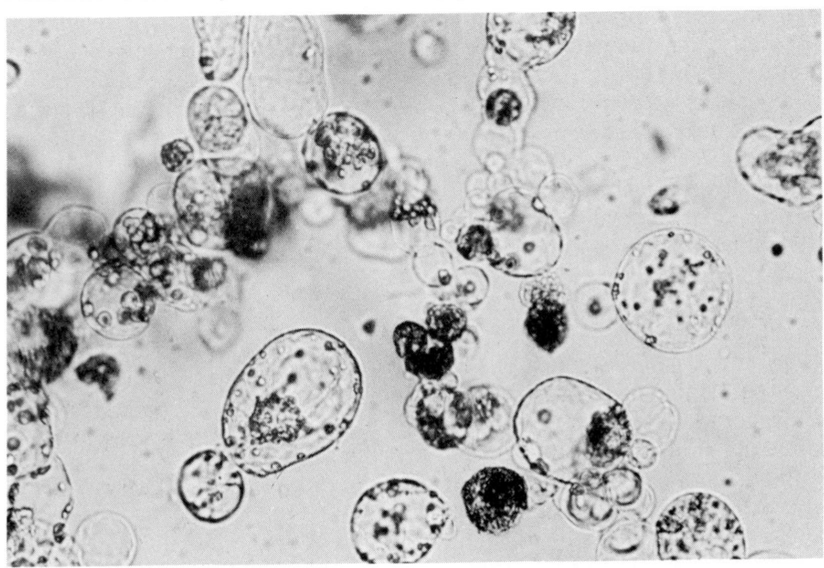

Fig. 9.2 These are cells regenerating from protoplasts, some of which have naturally undergone fusion, in sugar beet.

Moving on to one of the so-called 'shot-gun' techniques, it has been successfully demonstrated that by irradiating pollen of one variety of *Nicotiana rustica* immediately prior to fertilization, only a small portion of the paternal genome is expressed when hybridizing with another distinct variety of the same species (Jinks *et al.*, 1981). While the underlying genetic mechanism is by no means certain, the same phenomenon had also been reported when using pollen irradiation before making interspecific crosses, also in the genus *Nicotiana* (Pandey, 1978). The technique has now been used to transfer alien characteristics into an otherwise normal barley cultivar by Powell and his co-workers (1983) who are aiming to transfer in a systematic manner a large number of different sources of mildew resistance. Their studies may also lead them to investigate how the technique affects the potential for interspecific and intergeneric transfer of genetic information to *Hordeum*, which has often not been successsful using more conventional wide-crossing techniques.

Whether or not the techniques just described will yield results for the plant breeder in the immediate future remains a subject for debate. Nevertheless genetic manipulation, in the broad sense of transferring genetic material across the normal boundaries of species and genera by controlled wide crossing, will continue to play an important part in plant breeding as we have seen in Chapter 6 with maize. It will also continue to broaden the spectrum of utilized genetic resources. In order for progress to be made in this area, research needs to be expanded into embryo rescue operations aimed at obtaining viable seedlings from interspecific and intergeneric crosses. One of the methods that is widely used for this purpose consists of excision and culture of the undifferentiated hybrid embryo in a defined medium. Limited success has also been obtained by implanting the embryo on a nurse tissue of a normal endosperm or by culture of the hybrid ovule. In some cases, seed lethality is overcome by regeneration of plants by organogenesis from a callus originating from the hybrid embryo. Generally, this work has been well reviewed by Raghavan and others (1976, 1977, 1980), but it is worth looking more specifically at success stories illustrating broad germplasm utilization in this field. Wide crosses have improved dramatically through the use of embryo culture. This applies to several legume species and genera, and particularly to the cereals. Successes have included raising seedlings from *Triticum aestivum* × *Aegilops speltoides* hybrids (Chueca *et al.*, 1977), various *Triticum durum* × *Secale cereale* hybrids (Bajaj *et al.*, 1978), *Hordeum vulgare* × *Secale cereale* (Bajaj *et al.*, 1980) and *Aegilops squarrosa* × *Triticum boeoticum* (Gill *et al.*, 1981).

Transplantation or implantation of an embryo from its original hybrid endosperm environment to a normal endosperm may be employed to overcome the limitations of any artificial growth medium. Despite many technical difficulties, this technique may offer some promise as illustrated by Williams and De Lautour (1980) for pasture legumes.

One crop plant in which ovule culture has been successfully employed to rescue hybrid embryos is cotton, where tetraploid and diploid species have been crossed, their ovules cultured and grown to maturity (Stewart and Hsu, 1978).

Further research is needed in the future to identify the correct isolation conditions and nutrient media for successful mass isolation and culture of hybrid embryos and regeneration of seedling plants in quantity in a broader spectrum of crop plants.

Prospects

In this introduction to plant genetic resources, we have attempted to give the reader some indication of the whole range of activities which this field now encompasses. Only a few years ago little was appreciated about the magnitude of the problem. Fortunately many scientists and governments have woken up to the problem of genetic conservation before it is too late. Collecting missions have accomplished a great deal in capturing these precious resources before they disappear forever. The network of genebanks is growing, with the increasing evaluation and exploitation of germplasm in plant breeding programmes. What we do not want to see is plant genetic resources being regarded as museum collections. They must be used by both plant breeders and genetic engineers for the benefit of mankind.

In the main, we are optimistic about the prospects for plant genetic conservation activities throughout the world. A body of trained personnel is undertaking useful work in many countries, particularly in the 'gene-rich' countries of the developing world. Our main concern is the politicization of plant genetic resources, which as we outlined earlier, has tended to distract some scientists and politicians from the actual problems at hand. Undoubtedly such disputes will be settled, and it is our hope that this can be achieved without jeopardizing all the valuable work which has been undertaken to date.

References

Åberg, E. (1940). The taxonomy and phylogeny of *Hordeum* L. section *Cerealia* Ands. with special reference to Tibetan barleys. *Symbolae Botanicae Upsaliensis*, **4**, 1–156.

Allard, R. W. (1970). Population structure and sampling methods. In *Genetic Resources in Plants – Their Exploration and Conservation*, Frankel, O. H. and Bennett, E. (Eds). IBP Handbook No. 11. Blackwell Scientific Publications, Oxford and Edinburgh.

Allen, J. B. (1984). Strategies and methods for collecting *Theobroma*. *Plant Genetic Resources Newsletter*, **57**, 8–14.

Anderson, E. (1969). *Plants, Man and Life*. University of California Press, Berkeley and Los Angeles.

Arasu, N. T. and Rajanaidu, N. (1975). Conservation and utilization of genetic resources in the oil palm (*Elaeis guineensis*). In *South East Asian Plant Genetic Resources*, Williams, J. T., Lamoureux, C. H. and Soetjipto, N. W. (Eds). IBPGR-BIOTROP-LIPI, Bogor.

Arnold, M. H. (1963). The control of bacterial blight in rain grown cotton. I. Breeding for resistance in African upland varieties. *Journal of Agricultural Science, Cambridge*, **60**, 415–27.

Ashton, P. S. (1976). An approach to the study of breeding systems, population structure and taxonomy of tropical trees. In *Tropical Trees*, Burley, J. and Styles, B. T. (Eds). Linnean Society Symposium Series, 2. Academic Press, London.

Ayad, W. G. (1983). A study on the problem of the origin and evolution of cultivated barley species-group and its wild relatives. *Ph.D thesis*, University of Birmingham, England.

Ayad, W. G., Croston, R. P. and Khawlani, M. A. A. (1980). Collecting in the Yemen Arab Republic II. *Plant Genetic Resources Newsletter*, **43**, 25–7.

Bajaj, Y. P. S., Gill, K. S. and Sandha, G. S. (1978). Some factors enhancing the *in vitro* production of hexaploid *Triticale* (*Triticum durum* × *Secale cereale*). *Crop Improvement*, **5**, 62–72.

Bajaj, Y. P. S., Verma, M. M. and Dhanju, M. S. (1980). Barley × rye hybrids (*Hordecale*) through embryo culture. *Current Science*, **49**, 362–3.

Bayliss, M. W. (1980). Chromosomal variation in plant tissue cultures. In *International Review of Cytology, Supplement IIA: Perspectives in Plant Cell and Tissue Culture*, Vassil, K. (Ed.). Academic Press, New York.

Beadle, G. W. (1980). The ancestry of corn. *Scientific American*, **242**, 96–103.

Bennett, E. (1970). Tactics of plant exploration. In *Genetic Resources in Plants – Their Exploration and Conservation*, Frankel, O. H. and Bennett, E. (Eds). IBP Handbook No. 11. Blackwell Scientific Publications, Oxford and Edinburgh.

Ben-Ze'ev, N. and Zohary, D. (1973). Species relationships in the genus *Pisum* L. *Israel Journal of Botany*, **22**, 73–91.

Bevan, M., Flavell, R. B. and Chilton, M. D. (1983). A chimaeric antibiotic resistance gene as a selectable marker for plant cell transformation. *Nature*, **304**, 184–7.

Blixt, S. (1978). Descriptive list of genes for *Pisum*. *Pisum Newsletter*, **10**, 80–101.

Blixt, S. (1984). Application of computers to genebanks and breeding programmes. In *Crop Breeding: A Contemporary Basis*, Vose, P. B. and Blixt, S. G. (Eds). Pergamon Press, Oxford.

Blixt, S. and Williams, J. T. (Eds). (1982). *Documentation of genetic resources: a model*. IBPGR Secretariat, Rome.

Bosemark, N. O. (1979). Genetic poverty of the sugar beet in Europe. In *Broadening the Genetic Base of Crops*, Zeven, A. C. and Van Harten, A. M. (Eds). Pudoc, Wageningen.

Bowden, W. M. (1959). The taxonomy and nomenclature of the wheats, barleys and ryes and their wild relatives. *Canadian Journal of Botany*, **37**, 657—84.

Brazier, J. D., Hughes, J. F. and Tabb, C. B. (1976). Exploitation of natural tropical forest resources and the need for genetic and ecological conservation. In *Tropical Trees*, Burley, J. and Styles, B. T. (Eds). Linnean Society Symposium Series, 2. Academic Press, London.

Burley, J. (1976). Genetic systems and genetic conservation of tropical pines. In *Tropical Trees*, Burley, J. and Styles, B. T. (Eds). Linnean Society Symposium Series, 2. Academic Press, London.

Camadro, E. and Peloquin, S. J. (1980). The occurrence and frequency of 2n pollen in three diploid Solanums from northwest Argentina. *Theoretical and Applied Genetics*, **56**, 11–15.

Cambel, H. and Braidwood, R. J. (1970). An early farming village in Turkey. *Scientific American*, **223**, 50–56.

Centro Internacional de Agricultura Tropical (1981). *Annual Report 1981*. CIAT, Cali, Colombia.

Centro Internacional de la Papa (1978). Improved potato storage. *CIP Circular*, 6, no. 11.

Chang, T. T. (1976). *Manual on Genetic Conservation of Rice Germplasm for Evaluation and Utilization*. International Rice Research Institute, Los Baños, Philippines.

Chang, T. T. (1977). The origin, evolution, cultivation, dissemination and diversification of Asian and African rices. *Euphytica*, **25**, 425–41.

Chang, T. T., Marciano, A. P. and Loresto, G. C. (1977). Morpho-agronomic variousness and economic potentials of *Oryza glaberrima* and wild species in the genus *Oryza*. In *Report of Meeting on African Rice Species*. IRAT-ORSTOM, Paris.

Chang, T. T., Ou, S. H., Pathak, M. D., Ling, K. C. and Kauffman, H. E. (1975). The search for disease and insect resistance in rice germplasm. In *Crop Genetic Resources for Today and Tomorrow*, Frankel, O. H. and Hawkes, J. G. (Eds). Cambridge University Press, Cambridge.

Chang, T. T., Sharma, S. D., Adair, C. R. and Perez, A. T. (1972). *Manual for field collectors of rice*. International Rice Research Institute, Los Baños, Philippines.

Chueca, M., Cauderon, Y. and Tempe, J. (1977). Technique d'obtention d'hybrides blé tendre × *Aegilops* par culture *in vitro* d'embryons immatures. *Anneles de l'Amélioration des Plantes*, **27**, 539–47.

Coats, A. M. (1969). *The Quest for Plants*. Studio Vista, London.

Cocking, E. C. and Riley, R. (1981). Application of tissue culture and somatic hybridisation to plant improvement. In *Plant Breeding II*. Frey, K. J. (Ed.). Iowa State University Press, Iowa.

Conard, N., Asch, D. L., Asch, N. B., Elmore, D., Gove, H., Rubin, M., Brown, J. A., Wiant, M. D., Farnsworth, K. B. and Cook, T. G. (1984). Accelerator radiocarbon dating of evidence for prehistoric horticulture in Illinois. *Nature*, **308**, 443–6.

Consultative Group on International Agricultural Research (1982). *Report on the Consultative Group and the international agricultural research it supports*. CGIAR Secretariat, Washington D.C.

Cribb, P. J. (1972). Studies on the origin of *Solanum tuberosum* L. subsp. *andigena* (Juz. et Buk.) Hawkes – the cultivated tetraploid potato of South America. *Ph.D. Thesis*, University of Birmingham, England.

Crisp, P. and Ford-Lloyd, B. (1981). A different approach to vegetable germplasm collection. *Plant Genetic Resources Newsletter*, **48**, 11–12.

Cromarty, A. S., Ellis, R. H. and Roberts, E. H. (1982). *The design of seed storage facilities for genetic conservation*. IBPGR, Rome.

Cruden, R. W. (1977). Pollen-ovule ratios: a conservative indicator of breeding systems in flowering plants. *Evolution*, **31**, 32–46.

D'Amato, F. (1978). Chromosome number variations in cultured cells and regenerated plants. In *Frontiers of Plant Tissue Culture*. Thorpe, T. A. (Ed.). International Association for Plant Tissue Culture, Calgary.

Damania, A. B. and Porceddu, E. (1982). Screening YAR barley for disease resistance. *Plant Genetic Resources Newsletter*, **48**, 2–4.

Damania, A. B., Jackson, M. T. and Porceddu, E. (1985). Variation in wheat and barley landraces from Nepal and the Yemen Arab Republic. *Zeitschrift für Pflanzenzüchtung*, **94**, 13–24.

Damania, A. B., Porceddu, E. and Jackson, M. T. (1983). A rapid method for the evaluation of variation in germplasm collections of cereals using polyacrylamide gel electrophoresis. *Euphytica*, **32**, 877–83.

Darlington, C. D. and Janaki Ammal, E. K. (1945). *Chromosome Atlas of Cultivated Plants*. Allen and Unwin, London.

De Candolle, A. (1882). *Origine des Plantes Cultivées*. Paris. (English translation, 1886, Kegan Paul.)

De Wet, J. M. J. (1975). Evolutionary dynamics of cereal domestication. *Bulletin of the Torrey Botanical Club*, **102**, 307–12.

De Wet, J. M. J. (1979). Principles of evolution and cereal domestication. In *Broadening the Genetic Base of Crops*, Zeven, A. C. and Van Harten, A. M. (Eds). Pudoc, Wageningen.

De Wet, J. M. J. and Harlan, J. R. (1975). Weeds and domesticates: evolution in the man-made habitat. *Economic Botany*, **29**, 99–107.

De Wet, J. M. J., Harlan, J. R. and Grant, C. A. (1971). Origin and evolution of teosinte (*Zea mexicana* (Schrad.) Kuntze). *Euphytica*, **20**, 255–65.

Dodds, J. H. and Roberts, L. W. (1982). *Experiments in Plant Tissue Culture*. Cambridge University Press, Cambridge.

Dunning, A. (1983). Rhizomania – a threat to the UK crop? *British Sugar Beet Review*, **51**, 53–4.

Ellis, R. H. and Wetzel, M. (1983). Recent developments on applying sequential analysis to genebank and viability monitoring tests. *Plant Genetic Resources Newsletter*, **55**, 2–15.

Engelbrecht, T. H. (1916). Uber die Entstehung einiger feldmassig angebauter Kulturpflanzen. *Geographische Zeitschrift*, 22, 328–35.

Farnsworth, N. R. and Bingel, A. S. (1977). Problems and prospects of discovering new drugs from higher plants by pharmacological screening. In *New Natural Products and Plant Drugs with Pharmacological, Biological or Therapeutical Activity*, Wagner, H. and Wolff, P. (Eds). Springer-Verlag, Berlin.

Feldman, M. (1976). Wheats. In *Evolution of Crop Plants*, Simmonds, N. W. (Ed.). Longman, London and New York.

Ferwerda, F. P. (1976). Coffees. In *Evolution of Crop Plants*, Simmonds, N. W. (Ed.). Longman, London and New York.

Fischbeck, G. (1981). The usefulness of gene banks – perspectives for the breeding of plants. In *UPOV Symposium: The Use of Genetic Resources in the Plant Kingdom.*

Flavell, R. B. and Mathias, R. (1984). Prospects for transforming monocot crop plants. *Nature*, **307**, 108–9.

Food and Agriculture Organization (1976). *Production Yearbook*. Vol. 30, Rome.

Food and Agriculture Organization (1980). *Production Yearbook* 1979. Vol. 33, Rome.

Food and Agriculture Organization (1981). *Report of the Fifth Session of the FAO Panel of Experts on Forest Gene Resources, Rome*, December 1981. FAO, Rome.

Ford-Lloyd, B. V. and Williams, J. T. (1975). A revision of *Beta* section *Vulgares* (Chenopodiaceae), with new light on the origin of cultivated beets. *Botanical Journal of the Linnean Society*, **71**, 89–102.

Frankel, O. H. (1970). Genetic conservation in perspective. In *Genetic Resources in Plants: Their Exploration and Conservation*, Frankel, O. H. and Bennett, E. (Eds). IBP Handbook No. 11. Blackwell Scientific Publications, Oxford and Edinburgh.

Frankel, O. H. (1981). General principles of germplasm regeneration. In *Report of the FAO/UNEP/IBPGR International Conference on Crop Genetic Resources*. FAO, Rome.

Frankel, O. H. (Ed.) (1973). *Survey of crop genetic resources in their centres of diversity: first report*. FAO/IBP, Rome.

Frankel, O. H. and Bennett, E. (Eds) (1970). *Genetic Resources in Plants – Their Exploration and Conservation*. IBP Handbook No. 11. Blackwell Scientific Publications, Oxford and Edinburgh.

Frankel, O. H. and Hawkes, J. G. (Eds) (1975). *Crop Genetic Resources for Today and Tomorrow*. Cambridge University Press, Cambridge.

Ganeshaiah, K. N. and Uma Shaanker, R. (1982). Evolution of reproductive behaviour in the genus *Eleusine*. *Euphytica*, **31**, 397–404.

Geary, T. (1970). Direction and distance of pine pollen dispersal and seed orchard location on the Copperbelt of Zambia. *Rhodesian Journal of Agricultural Research*, **8**, 123–30.

George, E. F. and Sherrington, P. D. (1984). *Plant Propagation by Tissue Culture: Handbook and Directory of Commercial Laboratories*. Exegetics, Basingstoke.

Giessen, J. E., Hoffman, W. and Schottenloher, R. (1956). Die Gersten Athiopiens und Erythraas. *Zeitschrift für Pflanzenzüchtung*, 35, 377–440.

Gill, B. S., Waines, J. G. and Sharma, H. C. (1981). Endosperm abortion and the production of viable *Aegilops squarrosa* × *Triticum boeoticum* hybrids by embryo culture. *Plant Science Letters*, **23**, 181–7.

Godin, V. J. and Spensley, P. C. (1971). *Oils and Oilseeds*. No. 1, Crop and Products Digest. Tropical Products Institute, London.

Gorman, C. F. (1969). Hoabinthian: a pebble tool complex with early plant associations in south-east Asia. *Science*, **163**, 671–3.

Gregor, J. W. (1933). The ecotype concept in relation to the registration of crop plants. *Annals of Applied Biology*, **20**, 205–19.

Guldager, P. (1975). *Ex situ* conservation stands in the tropics. In *Methodology of Conservation of Forest Genetic Resources*, Roche, L. (Ed.). FAO, Rome.

Gupta, R., Sethi, K. L., Kazim, M. and Gupta, P. C. (1980). Opium poppy collection in Rajasthan. *Plant Genetic Resources Newsletter*, **43**, 20–4.

Harlan, H. V. (1957). *One Man's Life with Barley*. Exposition Press, New York.

Harlan, J. R. (1951). Anatomy of gene centres. *American Naturalist*, **85**, 97–103.

Harlan, J. R. (1966). Plant introduction and biosystematics. In *Plant Breeding*, Frey, K. J. (Ed.). Iowa State University Press, Ames.

Harlan, J. R. (1971a). Agricultural origins: centers and non-centers. *Science*, **174**, 468–74.

Harlan, J. R. (1971b). On the origin of barley: a second look. In *Barley Genetics II*, Nilan, R. A. (Ed.). Washington University Press, Washington.

Harlan, J. R. (1976). Genetic resources in wild relatives of crops. *Crop Science*, **16**, 329–33.

Harlan, J. R. (1979). On the origin of barley. In *Barley*, Agricultural Handbook no. 338. USDA, Washington.

Harlan, J. R. and De Wet, J. M. J. (1971). Toward a rational classification of cultivated plants. *Taxon*, **20**, 509–17.

Harlan, J. R. and De Wet, J. M. J. (1977). Pathway of genetic transfer from *Tripsacum* to *Zea mays*. *Proceedings of the National Academy of Sciences, USA*. **74**, 3494–7.

Harlan, J. R. and Zohary, D. (1966). Distribution of wild wheats and barley. *Science*, **13**, 1074–80.

Harrington, J. F. (1963). Practical advice and instructions on seed storage. *Proceedings of the International Seed Testing Association*, **28**, 989–94.

Harrington, J. F. (1972). Seed storage and longevity. In *Seed Biology*, Kozlowski, T. T. (Ed.). Academic Press, New York and London.

Harris, D. R. (1969). Agricultural systems, ecosystems and the origins of agriculture. In *The Domestication and Exploitation of Plants and Animals*, Ucko, P. J. and Dimbleby, G. W. (Eds). Duckworth, London.

Hawkes, J. G. (1962). The origin of *Solanum juzepczukii* Buk. and *S. curtilobum* Juz. et Buk. *Zeitschrift für Pflanzenzüchtung*, **47**, 1–14.

Hawkes, J. G. (1969). The ecological background of plant domestication. In *The Domestication and Exploitation of Plants and Animals*, Ucko, P. J. and Dimbleby, G. W. (Eds). Duckworth, London.

Hawkes, J. G. (1973). Potato genetic erosion survey – preliminary report. In *Germplasm Exploration and Taxonomy of Potatoes*. Report of the Planning Conference held at the International Potato Center, Lima, Peru, January 1973.

Hawkes, J. G. (1978). Biosystematics of the potato. In *The Potato Crop*, Harris, P. M. (Ed.). Chapman and Hall, London.

Hawkes, J. G. (1980). *Crop Genetic Resources Field Collection Manual*. IBPGR/ EUCARPIA, 37pp.

Hawkes, J. G. (1982). Genetic conservation of 'recalcitrant species' – an overview. In *Crop Genetic Resources: The Conservation of Difficult Material*, Withers, L. and Williams, J. T. (Eds). IUBS/IBPGR, Paris.

Hawkes, J. G. (1983).*The Diversity of Crop Plants*. Harvard University Press, Cambridge, Massachusetts and London.

Heijbroek, W., Roelands, A. J. and De Jong, J. H. (1983). Transfer of resistance to beet cyst nematode from *Beta patellaris* to sugar beet. *Euphytica*, **32**, 287–98.

Henshaw, G. G. (1984). Tissue culture for disease elimination and germplasm storage. In *Crop Breeding: A Contemporary Basis*, Vose, P. B. and Blixt, S. G. (Eds). Pergamon Press, Oxford.

Ho, P. (1969). The loess and the origin of Chinese agriculture. *American Historical Review*, **75**, 1–36.

Huaman, Z., Hawkes, J. G. and Rowe, P. R. (1980). *Solanum ajanhuiri*: An important diploid potato cultivated in the Andean altiplano. *Economic Botany*, **34**, 335–43.

Huaman, Z., Hawkes, J. G. and Rowe, P. R. (1982). A biosystematic study on the origin of the cultivated diploid potato *S. ajanhuiri* Juz. *et* Buk. *Euphytica*, **31**, 665–76.

Iltis, H. H., Doebley, J. F., Guzman, M. R. and Pazy, B. (1979). *Zea diploperennis* (Gramineae): a new teosinte from Mexico. *Science*, **203**, 186–8.

International Board for Plant Genetic Resources (1976). *Report of IBPGR Working Group on Engineering, Design and Cost Aspects of Long-Term Seed Storage Facilities*. IBPGR, Rome.

International Board for Plant Genetic Resources (1981). *Annual Report*. IBPGR, Rome.

International Board for Plant Genetic Resources (1981). *Genetic Resources of Tomatoes and wild Relatives*, by Esquinas-Alcazar, J. T. IBPGR, Rome.

International Board for Plant Genetic Resources (1982). *Descriptor List for* Allium, by Astley, D., Innes, N. L. and van der Meer Q.P. IBPGR, Rome.

International Board for Plant Genetic Resources (1983a). *Annual Report*. IBPGR, Rome.

International Board for Plant Genetic Resources (1983b). *Genetic resources of* Capsicum. IBPGR, Rome.

International Union for Conservation of Nature and Natural Resources (1980). *World Conservation Strategy*. Switzerland.

Jackson, M. T., Hawkes, J. G. and Rowe, P. R. (1977). The nature of *Solanum* × *chaucha* Juz. *et* Buk., a triploid cultivated potato from the South American Andes. *Euphytica*, **26**, 775–83.

Jackson, M. T., Hawkes, J. G. and Rowe, P. R. (1980). An ethnobotanical field study of primitive potato varieties in Peru. *Euphytica*, **29**, 107–13.

Jennings, D. L. (1976). Cassava. In *Evolution of Crop Plants*, Simmonds, N. W. (Ed.). Longman, London and New York.

Jinks, J. L., Caligari, P. D. S. and Ingram, N. R. (1981). Gene transfer in *Nicotiana rustica* using irradiated pollen. *Nature*, **291**, 586–8.

Johnson, B. L. (1975). Identification of the apparent B-genome donor of wheat. *Canadian Journal of Genetics and Cytology*, **17**, 21–9.

Jones, A. (1980). Sweet potato. In *Hybridization of Crop Plants*, Fehr, W. R. and Hadley, H. H. (Eds). American Society of Agronomy and Crop Science Society of America, Madison, Wisconsin.

Jones, H. A. and Mann, L. K. (1963). *Onions and their Allies: Botany, Cultivation and Utilization*. Interscience, New York.

Kawano, K. (1980). Cassava. In *Hybridization of Crop Plants*, Fehr, W. R. and Hadley, H. H. (Eds). American Society of Agronomy and Crop Science Society of America, Madison, Wisconsin.

Keefe, P. D. and Henshaw, G. G. (1984). A note on the multiple role of artificial nucleation of the suspending medium during two-step cryopreservation procedures. *Cryo-letters*, **5**, 71–8.

Kemp, R. H. (1975). Central American pines. In *Methodology of Conservation of Forest Genetic Resources*, Roche, L. (Ed.). FAO, Rome.

Kemp, R. H., Roche, L. and Willan, R. L. (1976). Current activities and problems in the exploration and conservation of tropical forest gene resources. In *Tropical Trees*, Burley, J. and Styles, B. T. (Eds). Linnean Society Symposium Series, 2. Academic Press, London.

Khush, G. S. (1977). Disease and insect resistance in rice. *Advances in Agronomy*, **29**, 265–341.

King, M. W. and Roberts, E. H. (1980). Maintenance of recalcitrant seeds in storage. In *Recalcitrant Crop Seeds*, Chin, H. F. and Roberts, E. H. (Eds). Tropical Press, Kuala Lumpur.

Kingdon-Ward, F. (1950). Does wild tea exist? *Nature*, **165**, 297.

Ladizinsky, G. (1979). The origin of the lentil and its wild genepool.*Euphytica*, **28**, 179–87.

Ladizinsky, G. and Adler, A. (1976). Genetic relationships among the annual species of *Cicer* L. *Theoretical and Applied Genetics*, **48**, 197–203.

Larsen, E. and Cromer, D. A. N. (1970). Exploration, evaluation, utilization and conservation of eucalypt gene resources. In *Genetic Resources in Plants – Their Exploration and Conservation*, Frankel, O. H. and Bennett, E. (Eds). IBP Handbook No. 11. Blackwell Scientific Publications, Oxford and Edinburgh.

Law, C. N., Snape, J. W. and Worland, A. J. (1981). Intraspecific chromosome manipulation. *Philosophical Transactions of the Royal Society of London*, B, **292**, 509–18.

Lehmann, C. O., Nover, I. and Scholz, F. (1976). The Gatersleben barley collection and its evaluation. In *Barley Genetics III*, Gaul, H. (Ed.). Theimig, Garching, Munich.

Leron Robbins, M. and Whitwood, W. N. (1973). Deep cold treatment of seeds: effect on germination and on callus production from excised cotyledons. *Horticultural Research*, **13**, 137–41.

MacNeish, R. S. (1964). Ancient Mesoamerican civilization. *Science*, **143**, 531–7.

Mangelsdorf, P. C. (1965). The evolution of maize. In *Essays on Crop Plant Evolution*, Hutchinson, J. (Ed.). Cambridge University Press, Cambridge.

Marshall, D. R. and Brown, A. H. D. (1975). Optimum sampling strategies in genetic conservation. In *Crop Genetic Resources for Today and Tomorrow*, Frankel, O. H. and Hawkes, J. G. (Eds). Cambridge University Press, Cambridge.

Martin, F. W. (1976). Cytogenetics and plant breeding of cassava. A review. *Plant Breeding Abstracts*, **46**, 909–16.

Martins, R. (1976). New archaeological techniques for the study of ancient root crops in Peru. *Ph.D. Thesis*, University of Birmingham, England.

Matthiolus, P. A. (1554). *Commentarii in sex libros Pedacii Dioscoridis*. Venice.

Mazur, P. (1966). Physical and chemical basis of injury in single-celled microorganisms subjected to freezing and thawing. In *Cryobiology*, Meryman, H. T. (Ed.). Academic Press, New York.

Mazur, P. (1976). Freezing and low temperature storage of living cells. In *Basic Aspects of Freeze Preservation of Mouse Strains*, Muhlbock, O. (Ed.). Gustav Fischer Verlag, Stuttgart.

Mooney, P. R. (1983). The law of the seed. *Development Dialogue*, **1–2**, 6–173.

Moseman, J. G., Nevo, E. and Zohary, D. (1983). Resistance of *Hordeum spontaneum* collected in Israel to infection with *Erysiphe graminis hordei*. *Crop Science*, **23**, 1115–19.

Mumford, P. M. and Grout, B. W. W. (1978). Germination and liquified nitrogen storage of cassava seed. *Annals of Botany*, **42**, 255–7.

Mumford, P. M. and Grout, B. (1979). Desiccation and low temperature (−196°C) tolerance of *Citrus limon* seed. *Seed Science and Technology*, **7**, 407–10.

Munck, L., Karlsson, K. E., Hagberg, A. and Eggum, B. O. (1970). Gene for improved nutritional value in barley seed protein. *Science*, **168**, 985.

Munerati, O. (1932). Sull incrocio della barbabietola coltivata con la beta selvaggia della costa adriatica. Industria Saccarifera Italiana, **25**, 303–4.

Murdock, G. P. (1959). Staple subsistence crops of Africa. *Geographical Review*, **50**, 521–40.

Ooi, S. C. and Rajanaidu, N. (1979). Establishment of oil palm genetic resources – theoretical and practical considerations. *Malaysian Applied Biology*, **8**, 15–28.

Pandey, K. K. (1978). Gametic gene transfer in *Nicotiana* by means of irradiated pollen. *Genetics*, **49**, 56–69.

Parnell, F. R., King, H. E. and Ruston, D. F. (1949). Jassid resistance and hairiness of the cotton plant. *Bulletin of Entomological Research*, **39**, 539–75.

Peat, J. E. and Brown, K. J. (1961). A record of cotton breeding for the Lake Province of Tanganika: seasons 1939–40 to 1957–58. *Empire Journal of Experimental Agriculture*, **29**, 119-35.

Peloquin, S. J. (1981). Chromosomal and cytoplasmic manipulations. In *Plant Breeding II*, Frey, K. J. (Ed.). Iowa State University Press, Iowa.

Pickersgill, B. (1981). Biosystematics of crop-weed complexes. *Kulturpflanze*, **29**, 377–88.

Pomeranz, Y. (1973). A review of proteins in barley, oats and buckwheat. *Cereal Science Today*, **18**, 310.

Pomeranz, Y. (1976). Proteins and amino acids of barley. In *Barley Genetics III*, Gaul, H. (Ed.). Theimig, Garching, Munich.

Poore, M. E. D. (1968). Studies in Malayan rain forest. 1. The forest on triassic sediments in Jengka Forest Reserve. *Journal of Ecology*, **56**, 143–96.

Portères, R. (1950). Vieilles agricultures de l'Afrique intertropical. *Agronomie Tropicale*, 5, 489–507.

Powell, P., Caligari, P. D. S. and Hayter, A. M. (1983). The use of pollen irradiation in barley breeding. *Theoretical and Applied Genetics*, **65**, 73–6.

Qualset, C. O. and Suneson, C. A. (1966). A barley gene-pool for use in breeding for resistance to the barley yellow dwarf virus disease. *Crop Science*, **6**, 302.

Quinn, A. A., Mok, D. W. S. and Peloquin, S. J. (1974). Distribution and significance of diplandroids among the diploid Solanums. *American Potato Journal*, **51**,16–21.

Raghavan, V. (1976). *Experimental Embryogenesis in Vascular Plants*. Academic Press, London.

Raghavan, V. (1977). Applied aspects of embryo culture. In *Applied and Fundamental Aspects of Plant Cell, Tissue and Organ Culture*, Reinert, J. and Bajaj, Y. P. S. (Eds). Springer-Verlag, Berlin.

Raghavan, V. (1980). Embryo culture. *International Review of Cytology Supplement*, **IIB**, 209–40.

Richardson, S. D. (1970). Gene pools in forestry. In *Genetic Resources in Plants – Their Exploration and Conservation*, Frankel, O. H. and Bennett, E. (Eds). IBP Handbook, No. 11. Blackwell Scientific Publications, Oxford and Edinburgh.

Rick, C. M. (1976). Tomatoes. In *Evolution of Crop Plants*, Simmonds, N. (Ed.). Longman, London and New York.

Riley, R. (1965). Cytogenetics and evolution of wheat. In *Essays on Crop Plant Evolution*, Hutchinson, J. (Ed.). Cambridge University Press, Cambridge.

Roberts, E. H. (1972). Storage environment and the control of viability. In *Seed Viability*, Roberts, E. H. (Ed.). Chapman and Hall, London.

Roberts, E. H. (1973). Predicting the storage life of seeds. *Seed Science and Technology*, **1**, 499–514.

Roberts, E. H. and King, M. W. (1982). Storage of recalcitrant seeds. In *Crop Genetic Resources: The Conservation of Difficult Material*, Withers, L. A. and Williams, J. T. (Eds). IUBS/IBPGR, Paris.

Roca, W. M., Bryan, J. E. and Roca, M. R. (1979). Tissue culture for the international transfer of potato genetic resources. *American Potato Journal*, **56**, 1–10.

Roche, L. R. (1975). Guidelines for the methodology of conservation of forest genetic resources. In *Methodology of Conservation of Forest Genetic Resources*, Roche, L. R. (Ed.). FAO, Rome.

Rogers, D. J. and Appan, S. G. (1973). Manihot, Manihotoides (*Euphorbiaceae*). Flora Neotropica, Monograph 13, New York.

Rogers, D. J. and Fleming, H. S. (1973). A monograph of *Manihot esculenta* with an explanation of the taximetric methods used. *Economic Botany*, **27**, 1—113.

Sakai, A. (1965). Survival of plant tissue at super-low temperatures. III. Relation between effective prefreezing temperatures and degree of frost hardiness. *Plant Physiology*, **40**, 882–87.

Sakai, A. and Nishiyama, Y. (1978). Cryopreservation of winter vegetative buds of hardy fruit trees in liquid nitrogen. *Horticultural Science*, **13**, 225–7.

Sakai, A. and Noshiro, M. (1975). Some factors contributing to the survival of crop seeds cooled to the temperature of liquid nitrogen. In *Crop Genetic Resources for Today and Tomorrow*, Frankel, O. H. and Hawkes, J. G. (Eds). Cambridge University Press, Cambridge.

Salunkhe, D. K., Bhonsle, K. I., Salunkhe, V. D. and Adsule, R. N. (1984). Anticancer agents of plant origin. *Critical Reviews in Plant Sciences*, **1**, 203–25.

Sarkar, P. and Stebbins, G. L. (1956). Morphological evidence for the origin of the B genome in wheat. *American Journal of Botany*, **43**, 297–304.

Satake, T. and Toriyama, K. (1979). Two extremely cool-tolerant varieties. *International Rice Research Newsletter*, **4**, 9–10.

Sauer, C. O. (1952). *Agricultural Origins and Dispersals*. American Geographical Society, New York.

Savitsky, H. (1975). Hybridization between *Beta vulgaris* and *B. procumbens* and transmission of nematode (*Heterodera schachtii*) resistance to sugar beet. *Canadian Journal of Genetics and Cytology*, **17**, 197–209.

Schiemann, E. (1932). Entstechung der Kulturpflanzen. *Handbuch der Vererbungswissenschaft*, **3**, 1–377.

Schmiediche, P. E., Hawkes, J. G. and Ochoa, C. M. (1982). The breeding of the cultivated potato species *Solanum* × *juzepczukii* and *S. curtilobum*. II. The resynthesis of *S. juzepczukii* and *S. curtilobum*. *Euphytica*, **31**, 695–707.

Shao, Q. (1981). The evolution of cultivated barley. In *Barley Genetics IV*. Edinburgh.

Sidhu, G. S., Khush, G. S. and Mew, T. W. (1978). Genetic analysis of bacterial blight resistance in 74 cultivars of rice, *Oryza sativa* L. *Theoretical and Applied Genetics*, **53**, 105–11.

Simmonds, N. W. (1962). Variability in crop plants, its use and conservation. *Biological Reviews*, **37**, 442–65.

Singh, I. D. (1980). Tea germplasm in India. *Plant Genetic Resources Newsletter*, **43**, 12–16.

Smartt, J. (1969). Evolution of American *Phaseolus* beans under domestication. In *Domestication and Exploitation of Plants and Animals*, Ucko, P. J. and Dimbleby, G. W. (Eds). Duckworth, London.

Smartt, J. (1978). The evolution of pulse crops. *Economic Botany*, **32**, 185–98.

Smartt, J. (1981). Genepools in *Phaseolus* and *Vigna* cultigens. *Euphytica*, **30**, 445–9.

Smartt, J. and Haq, N. (1972). Fertility and segregation of the amphidiploid *Phaseolus vulgaris* L. × *P. coccineus* L. and its behaviour in backcrosses. *Euphytica*, **21**, 496–501.

Smartt, J., Gregory, W. C. and Gregory, M. P. (1978). The genomes of *Arachis hypogaea*. 2. The implications in interspecific breeding. *Euphytica*, **27**, 677–80.

Soria, J. (1975). Recent cocoa collecting expeditions. In *Crop Genetic Resources for Today and Tomorrow*, Frankel, O. H. and Hawkes, J. G. (Eds). Cambridge University Press, Cambridge.

Stace, C. A. (1980). *Plant Taxonomy and Biosystematics*. Edward Arnold, London.

Stewart, J. M. and Hsu, C. L. (1978). Hybridization of diploid and tetraploid cottons through *in-ovulo* embryo-culture. *Journal of Heredity*, **69**, 404–8.

Sykes, J. T. (1975). Tree crops. In *Crop Genetic Resources for Today and Tomorrow*, Frankel, O. H. and Hawkes, J. G. (Eds). Cambridge University Press, Cambridge.

Thurston, H. D. and Lozano, J. C. (1968). Resistance to bacterial wilt of potatoes in Colombian clones of *Solanum phureja*. *American Potato Journal*, **45**, 51–5.

Toxopeus, H. and Van Sloten, D. H. (1981). The genetic resources of cruciferous crops. *Plant Genetic Resources Newsletter*, **47**, 3–8.

Tyler, B. F. (1979). Collections of forage grass in Europe. In *Broadening the Genetic Base of Crops*, Zeven, A. C. and Van Harten, A. M. (Eds). Pudoc, Wageningen.

Tyler, B. F., Chorlton, K. H. and Thomas, I. D. (1982). Collection and utilization of genetic resources of forage grasses and clover at the Welsh Plant Breeding Station. In *Report of the Welsh Plant Breeding Station for 1981*.

Uma Shaanker, R. and Ganeshaiah, K. N. (1980). Evolutionary significance of pollen to ovule ratio – a study in some pulse crops. *Current Science*, **49**, 244–5.

Valentine, R. C. (1978). Genetic blueprints for new plants. *Cereal Foods World*, **23**, 588–9.

Vavilov, N. I. (1926). Studies on the origins of cultivated plants. *Bulletin of Applied Botany and Plant Breeding*, **16**, 1–245.

Vavilov, N. I. (1951). The origin, variation, immunity and breeding of cultivated plants. *Chronica Botanica*, **13**, 1–366.

Villiers, T. A. (1974). Seed ageing: chromosome stability and extended viability of seeds stored fully imbibed. *Plant Physiology*, **53**, 875–8.

Wellhausen, E. J. (1978). Recent developments in maize breeding in the tropics. In *Maize Breeding and Genetics*, Walden, D. B. (Ed.). Wiley, New York.

Westcott, R. J. (1981a). Tissue culture storage of potato germplasm. 1. Minimal growth storage. *Potato Research*, **24**, 331–42.

Westcott, R. J. (1981b). Tissue culture storage of potato germplasm. 2. Use of growth retardants. *Potato Research*, **24**, 343–52.

Whitmore, T. C. (1972). In *Tree Flora of Malaya*, 1, Whitmore, T. C. (Ed.). Longman, Kuala Lumpur and London.

Wilkes, H. G. (1979). Mexico and Central America as a centre for the origin of agriculture and the evolution of maize. *Crop Improvement*, **6**, 1–18.

Wilkins, C. P. and Dodds, J. H. (1983). The application of tissue culture techniques to genetic conservation. *Science Progress, Oxford*, **68**, 259–84.

Wilkins, C. P., Bengochea, T. and Dodds, J. H. (1982). The use of *in vitro* methods for plant genetic conservation. *Outlook on Agriculture*, **11**, 67–72.

Williams, E. G. and De Lautour, G. (1980). The use of embryo culture with transplanted nurse endosperm for the production of interspecific hybrids in pasture legumes. *Botanical Gazette*, **141**, 252–7.

Williams, J. T. and Ford-Lloyd, B. V. (1975). Beet in Turkey. *Plant Genetic Resources Newsletter*, **31**, 3–6.

Williams, W. (1981). Methods of production of new varieties. *Philosophical Transactions of The Royal Society of London*, B, **292**, 421–30.

Willis, J. C. (1922). *Age and Area*. Cambridge University Press, Cambridge.

Witcombe, J. R. and Gilani, M. M. (1979). Variation in cereals from the Himalayas and the optimum strategy for sampling plant germplasm. *Journal of Applied Ecology*, **16**, 633–40.

Withers, L. A. (1980). *Tissue culture storage for genetic conservation*. Technical Report, IBPGR, Rome.

Withers, L. A. (1982). Storage of plant tissue cultures. In *Crop Genetic Resources: The Conservation of Difficult Material*, Withers, L. A. and Williams, J. T. (Eds). IUBS/IBPGR, Paris.

Yamashita, K. (1979). Origin and dispersion of wheats with special reference to peripheral diversity. *Seiken Ziho*, **27–28**, 48–58.

Yen, D. E. (1974). The sweet potato and Oceania. *B. P. Bishop Museum Bulletin*, 236, Honolulu.

Yen, D. E. (1976). Sweet potato. In *Evolution of Crop Plants*, Simmonds, N. W. (Ed.). Longman, London and New York.

Yngaard, F. (1983). A procedure for packing long-term storage seeds. *Plant Genetic Resources Newsletter*, **54**, 28–31.

Zeven, A. C. and De Wet, J. M. J. (1982). *Dictionary of Cultivated Plants and their Regions of Diversity*. Pudoc, Wageningen.

Zhukovsky, P. M. (1968). (New centres of origin and new gene centres of cultivated plants including specifically endemic microcentres of species closely allied to cultivated species.) *Botanicheskii Zhurnal*, 53, 430–60, *Plant Breeding Abstracts*, **38**, (38).

Zohary, D. (1959). Is *Hordeum agriocrithon* the ancestor of 6-rowed cultivated barley? *Evolution*, **13**, 279–80.

Zohary, D. (1970). Centres of diversity and centres of origin. In *Genetic Resources in Plants – Their Exploration and Conservation*, Frankel, O. H. and Bennett, E. (Eds). IBP Handbook, No. 11. Blackwell Scientific Publications, Oxford and Edinburgh.

Zohary, D., Harlan, J. R. and Vardi, A. (1969). The wild diploid progenitors of wheat and their breeding value. *Euphytica*, **18**, 58–65.

Appendix

Important germplasm collections throughout the world.

Genebank	*Cultivated forms*		*Wild relatives*
Wheat			
US Small Grains Collection, Beltsville Ag. Res. Center, Germplasm Resources Lab., Beltsville, Maryland 20705, USA.		36 710 (*Triticum*) 381 (*Aegilops*)	
N.I. Vavilov All-Union Institute of Plant Industry, 44 Herzen St., Leningrad 190000, USSR.		70 000 (39)	
Istituto del Germoplasma del CNR, Via G. Amendola 165/A, 70126 Bari, Italy		26 000 (27)	
Maize			
Inst. of Maize Res., Germplasm Dept., P.O. Box 89, Zemun 11081, Belgrade, Yugoslavia.		15 000	
N.I. Vavilov All-Union Institute of Plant Industry		15 084	
International Center for Maize and Wheat Improvement (CIMMYT), A.P. 6–641, Londres 40, Mexico D.F.		13 850 (3)	
Rice			
International Rice Research Institute (IRRI), P.O. Box 933, Manila, Philippines.	55 717		955
National Bureau of Plant Genetic Resources (NBPGR), Indian Agricultural Research Institute Campus, New Delhi 110012, India		33 833	
National Institute of Agricultural Sciences, Div. of Genetics, Soil Storage Laboratory, Kannondai 3.1.1, Yatabe-machi, Tsukuba-gun, Ibaraki-ken 300–21, Japan.	18 420		500

( ) = Number of species

<table>
<tr><th>Genebank</th><th>Cultivated forms</th><th>Wild relatives</th></tr>
<tr><td>Barley
U.S. Small Grains Collection, Beltsville.</td><td colspan="2">23 371</td></tr>
<tr><td>N.I. Vavilov All-Union Institute of Plant Industry.</td><td colspan="2">17 459</td></tr>
<tr><td>Nordic Genebank,Box 41, S–230 53 Alnarp, Sweden</td><td>11 900 (incl. mutants)</td><td>2 200</td></tr>
<tr><td>Potato
International Potato Center (CIP), Apartado 5969, Lima, Peru.</td><td>5 000+(9)</td><td>1 300(120)</td></tr>
<tr><td>Inter-Regional Potato Introduction Station, Sturgeon Bay, Wisconsin 54235, USA.</td><td>846</td><td>2 000 +</td></tr>
<tr><td>Commonwealth Potato Collection, Scottish Crop Research Institute, Pentlandfield, Roslin, Midlothian EH25 9RF, UK.</td><td colspan="2">6 000 +</td></tr>
<tr><td>Dutch-German Potato Collection, Institüt für Pflanzenbau und Saatgutforschung, FAL, Bundesalle 50, 33 Braunschweig, F.R. Germany.</td><td>635</td><td>1 735(67)</td></tr>
<tr><td>Cassava
International Center for Tropical Agriculture (CIAT), Apartado Aereo 6713, Cali, Colombia.</td><td colspan="2">2 450</td></tr>
<tr><td>International Institute for Tropical Agriculture (IITA), PMB 5320, Ojo Road, Ibadan, Nigeria.</td><td colspan="2">2 922</td></tr>
<tr><td>Sweet Potato
Asian Vegetable Research and Development Center (AVRDC), P.O. Box 42, Shanhua, Tainan 741, Taiwan.</td><td>1 200</td><td></td></tr>
<tr><td>Lembaga Pusat Penilitian Pertanian, P.O. Box 110, Bogor, Indonesia.</td><td>1 200</td><td></td></tr>
</table>

Genebank	*Cultivated forms*		*Wild relatives*
Food Legumes			
CIAT	*Phaseolus* spp.	22 000	
NBPGR	*Pisum, Vigna* etc.	6 300 +	
Genetic Resources Unit, International Crops Research Institute for Semi-Arid Tropics (ICRISAT), 1–11–256 Begumpet, Hyderabad 500016 a.p. India.	*Cicer, Cajanus, Arachis*	12 000	
International Center for Agricultural Research in Dry Areas (ICARDA), P.O. Box 5466, Aleppo, Syria.	*Lens, Cicer, Vicia*	12 000	
N.I. Vavilov All-Union Institute of Plant Industry	*Pisum, Vicia, Lupinus, Glycine, Phaseolus, Lens, Vigna* etc.	29 000	
Tomato			
AVRDC	4 320		342(5)
U.S.D.A. North Central Regional Plant Intro. St. (NC-7), Iowa State Univ., Ames, Iowa 50010, USA.	3 784		379(5)
Tomato Genetic Stock Center, Dept. of Vegetable Crops, University of California, Davis, California 95616, USA.	778		740(11)
Cocoa			
Cocoa Res. Unit, Fac. of Agric., Univ. of West Indies, St. Augustine. Trinidad.	762		
Cocoa Res. Inst. of Ghana, P.O. Box 8, Tafo-Akim, Ghana.		6 000 (15)	
Centro Agronómico Tropical de Investigación y Enseñanza (CATIE), Turrialba, Costa Rica.		2 000 (8+)	

Tea			
Tea Res. Inst. of East Africa, P.O. Box 91, Kericho, Kenya.	15 000		
Nat. Res. Inst. of Tea, Min. of Agric., Forestry and Fisheries, 2769 Kanaya, Shizuoka 428, Japan.	1 760		
Tocklai Exp. Sta., Jorhat, 785 008 Assam, India.	621		91(10)

INDEX

Index

Index